HKPIP

City Embellishment

城市装饰

香港理工国际出版社 主编

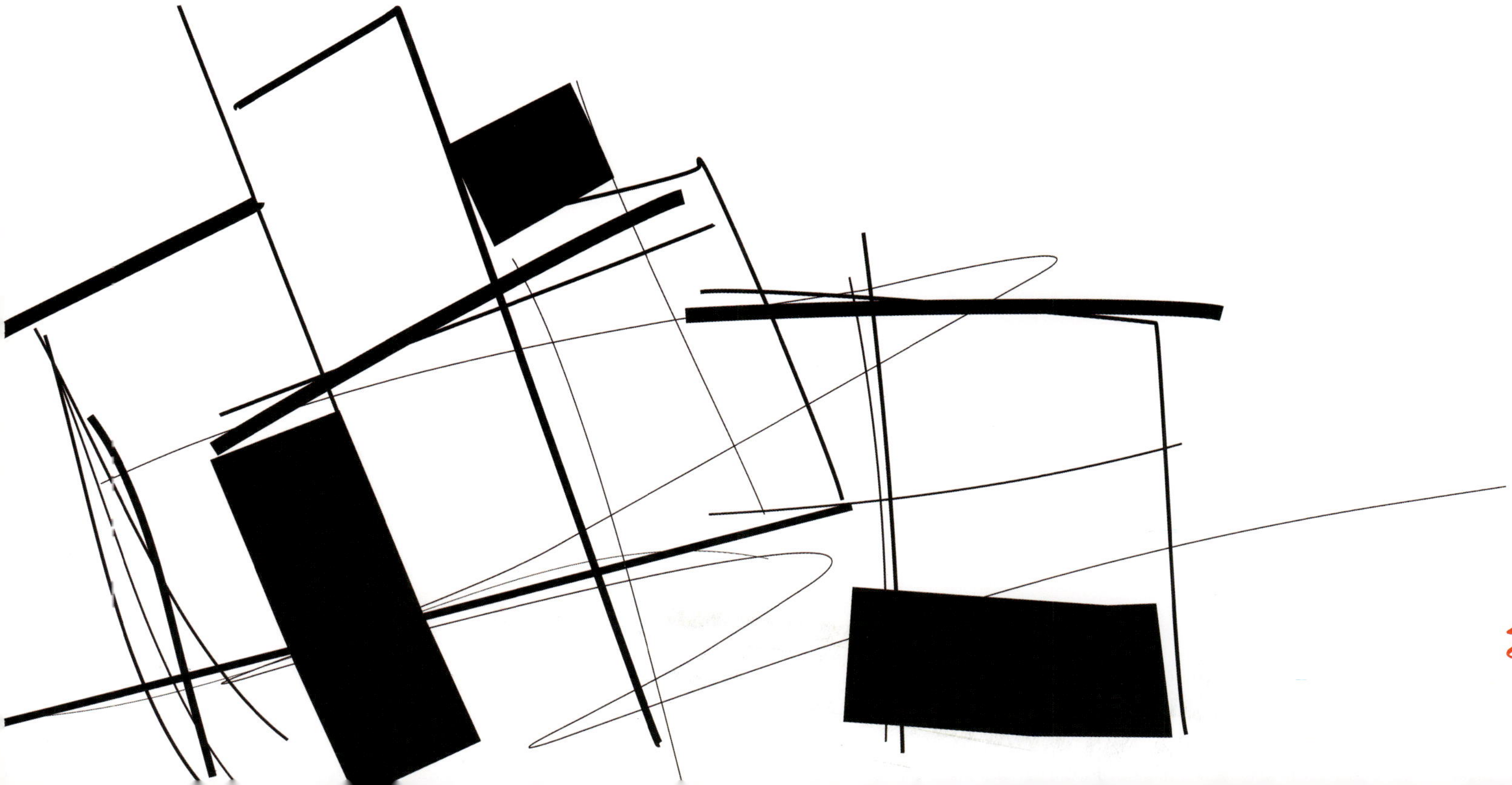

華中科技大學出版社
http://www.hustp.com
中国 · 武汉

Embellish C

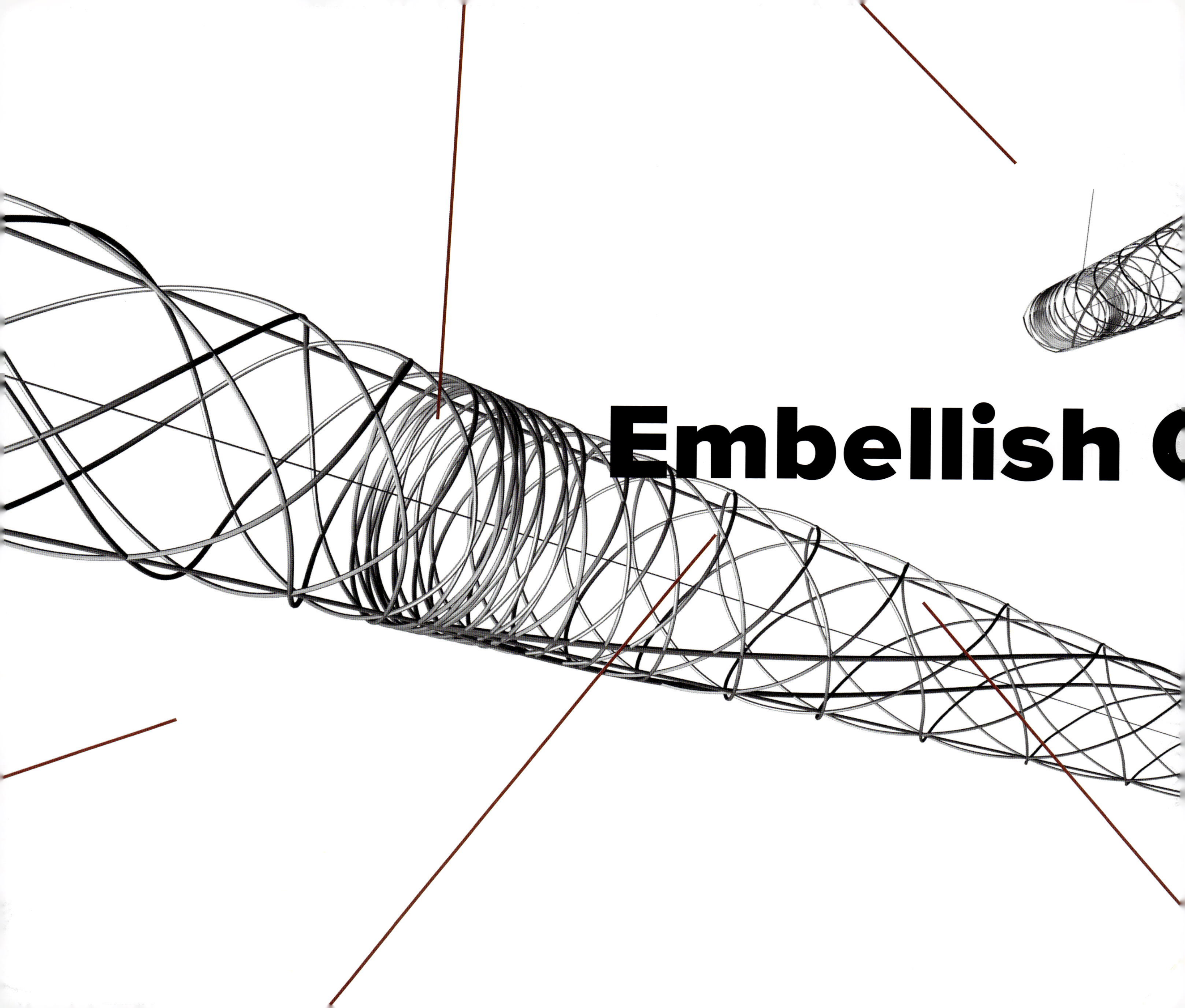

用艺术装点城市

ity with Arts

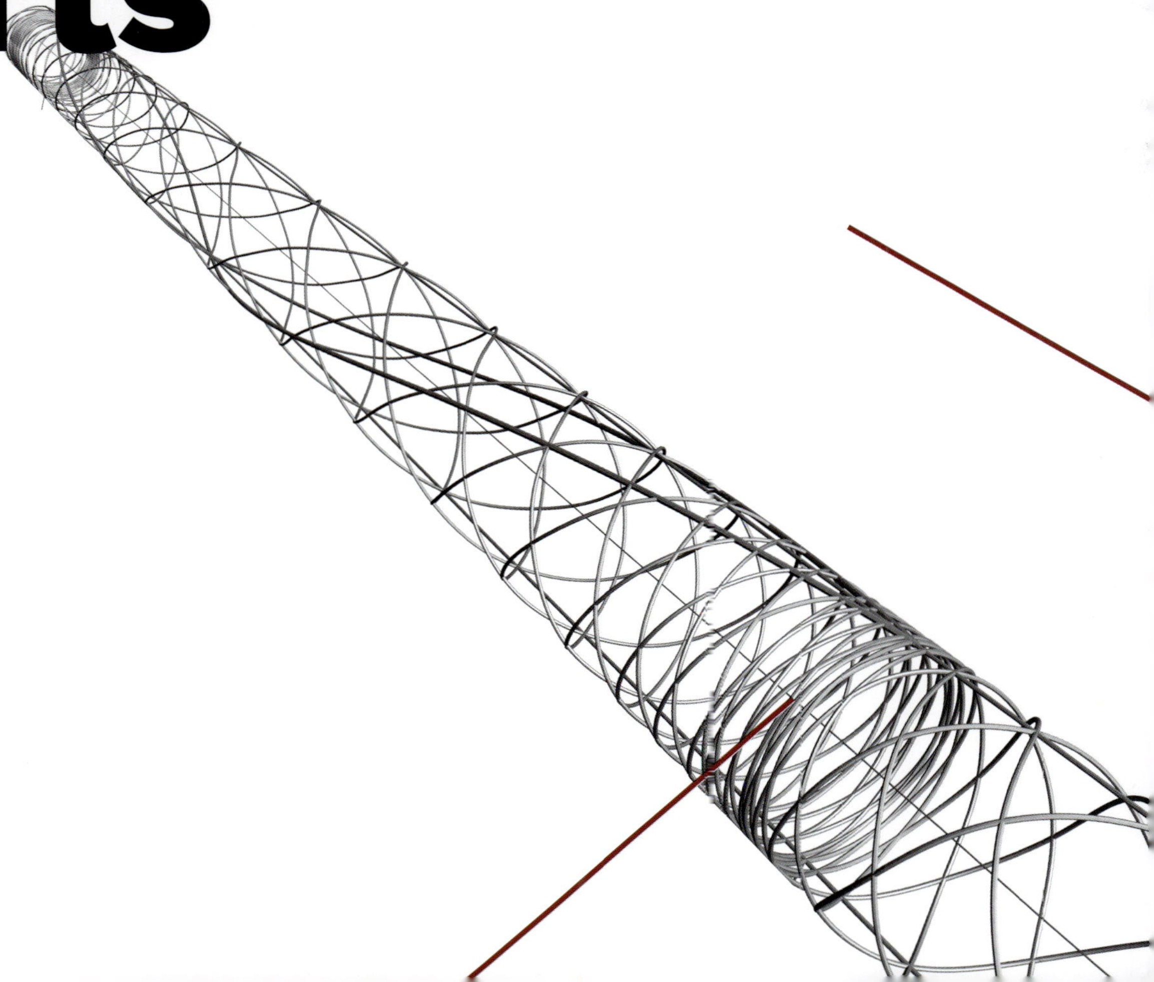

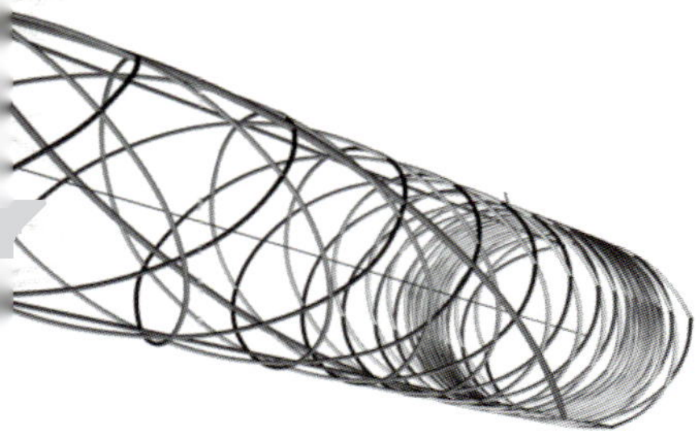

PREFACE

The beauty of a city is not in between the mountains and landscapes, neither in the brilliantly lit lights. Architectural design does not help to complete the beauty of itself, but to achieve the neighboring landscape, or in another way, the whole city, and even the economy of the country and its further development. At present, city embellishment and landscape design have integrated and aroused a great concern. Architecture and embellishment are all inspired from designers' observe and research in life.

City, as a place for people gathering, developed with the evolution of civilization. Districts have given the basic background to form the foremost sediment of civilization. Personification development of city civilization is the spatial expression of life. It's also an overall pattern of the living condition, behavior way, mental characteristics of the citizen and urban style. It's the soul of the city and a symbol of civilization. What's more, it's the reflection of its people's full value.

Economy is the image of the city, while culture is the spirit. Only to obtain image and spirit altogether, can a city be vibrant and dynamic, and to improve its outer and inner beauty by insistent spiritual civilization which will be rooted in every corner of the city. The overall appearance brings it not only elegance, but further boost the economic growth. Moreover, it's a gradual catalyst to an international reputation and status.

This book has collected 48 projects relating to city embellishment, which has included 12 landscape design, 15 interior design and 11 architectural design projects. All the projects have collected culture, history, art and inspiration as a whole. The aim of such projects is to provide a leisure space and landscape for the exchange of ideas, interactive entertainment and art appreciation of people. The publication of the book, while giving the designers inspiration, will also lead the reader to appreciate an elegant city embellishment, and then to create more fantastic works.

前言

城市之美不在乎于山水之间，也不在乎于霓虹灯的华丽初上。建筑设计远不止成就建筑本身那么简单，而是成就了它周边的所有建筑，所有氛围，抑或是说成就的是一座城市，乃至一个国家的经济状况和未来的发展之道。目前，城市装饰与景观设计已融为一体并受到普遍关注。建筑及装饰的设计都源自于设计师对生活的留心观察和悉心探索。

城市作为人类生活聚落的形态，伴随着文明的演进而发展。地域赋予了城市文化基本的底色，形成城市文化最初的积淀。城市文化的人格化发展，是人类生活的空间化表述，它是城市人群生存状况、行为方式、精神特征及城市风貌的总体形态，它是城市的灵魂，文明的标志，是属于这个城市人群的完整价值体现。

经济是城市的形，文化是城市的神，只有形神兼备，城市才能生机勃勃、充满活力，并用坚持不懈的精神文明建设提升自身的外在美和内在美，文明便在城市的各个角落生根发芽。城市的外貌带给它的不仅是华丽，而是进一步推进了经济的增长，以及逐渐提升其在国际上的名誉与地位。

本书共辑录了 48 个有关城市装饰的项目，其中包括景观设计 22 个，室内设计 15 个和建筑设计 11 个，是集文化、历史、艺术、灵感于一体的综合设计。为人们提供一个交流信息、娱乐互动、鉴赏艺术的休闲场所和城市景观，更有设计独特的室内装饰。藉此书的出版带给各位设计工作者及艺术爱好者们一个宏观的城市美感体验，在激发灵感的同时，为读者呈现出更多更好的艺术作品。

目录

CONTENTS

Architectural Embellishment

Landscape Embellishment
景观装饰

Landscape

Interior Decoration

caprabo

Architectural Embellishment

建筑装饰

La Roche-Sur-Yon Footbridge

永河畔拉罗什人行天桥

Project Location: France
Architectural Design: Bernard Tschumi Architects
Completion Year: 2009
Client: City of Roche Sur Yon
Photography: HAD
Building materials: Steel

项目地点：法国
建筑设计：伯纳德 • 屈米建筑公司
竣工时间：2009 年
开 发 商：法国拉罗什续尔雍市
摄　　影：HAD
建筑材料：钢材

A Tunnel to History and Reality 穿越历史与现实的时空隧道

Project Description

The extension of the TGV grain to La Roche-Sur-Yon, France, and nearby towns bordering the Atlantic marks not only an important moment for the modernization of the European at French train network, but also an occasion to initiate civic improvements. Linking the historic city founded by Napoleon ("the Pentagon") with new neighborhoods, this pedestrian bridge crosses above high-speed railway tracks, providing an important urban connection for the town. Conceived through joint collaboration between the fields of architecture and engineering, the bridge was designed by Bernard Tschumi and Hugh Dutton, with their respective teams in Paris and New York. The teams developed the design for La Roche-Sur-Yon as both a utilitarian vector of movement and a symbol of contemporary urban relationships. The intention of the designers was to demonstrate an integration of an original structural system with an architectural concept developed from urban scale research of neighborhood identity and carried through the express of the minutest details.

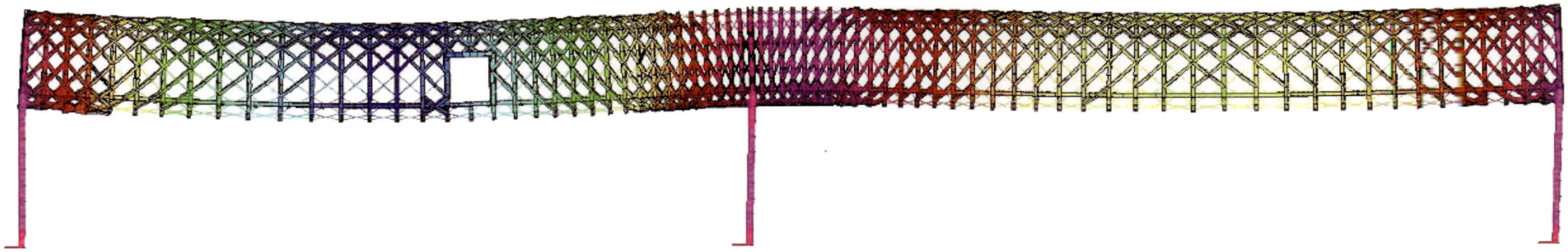

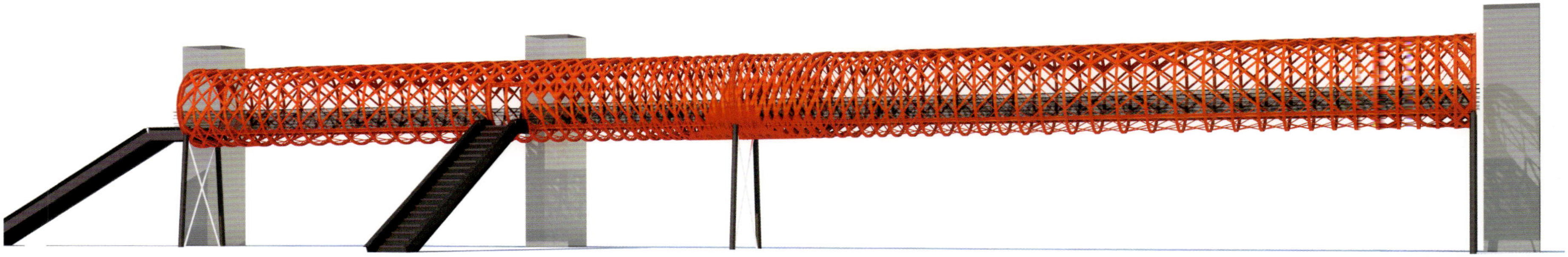

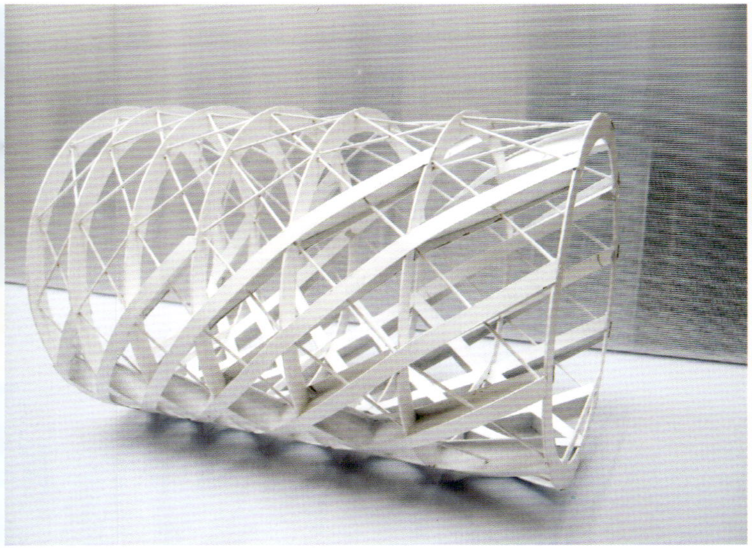

Design Details

It has been said that there is no architecture without movement. A pedestrian bridge is not just a static object, but represents a dynamic vector in both its usage and urban perception. The designers have sought to express this dynamic characteristic, as much through the structural system as through finishing materials (interlaced polycarbonate surfaces protect passengers from weather conditions, while lighting follows the rhythm of the structure). Even the bright red-orange color was chosen to emphasize the urban significance of the bridge as a pedestrian vector. The design of the new bridge uses the same language of a diagonal mesh, but in a tubular form, to create a complete cylindrical volume through which the users pass. Footbridges over railways require lateral protection for safety of both the users and the trains below. The complete volume provides a single structural solution that possesses the necessary inertia to span between the available support points as well as provide support for the required protective screens and a canopy cover.

项目描述

法国的高速列车向大西洋边界的永河畔拉罗什市及周边城镇的延伸，不仅标志着法国铁路在欧洲现代化建设中所迈开的重要一步，更是一个发起全民提高自身素质的好机会。这座新近落成的天桥宛若红色的弹簧一般，呈圆筒网状，以表现桥身的荷载与张力，同时为靠近西海岸的永河罗什镇创造一个全新而现代的建筑形象。由伯纳德 • 屈米和休达顿以及他们各自在巴黎和纽约的建筑团队设计，构想通过建筑学与工程学的结合，来共同完成了此项目。该项目的设计使它成为运动矢量和当代城市关系的一种象征。设计师的设计意图是从城市规模原始架构系统与建筑理念，从城市规模邻里身份的研究和通过微小的细节表达出来。

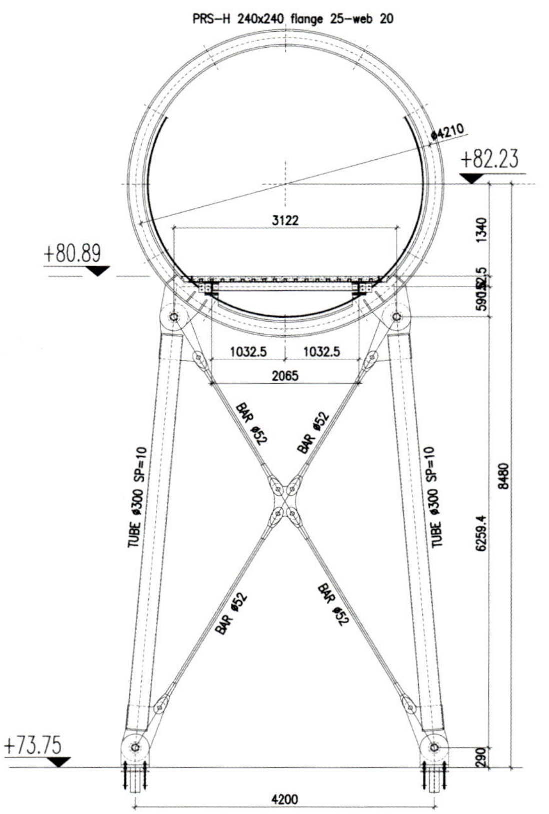

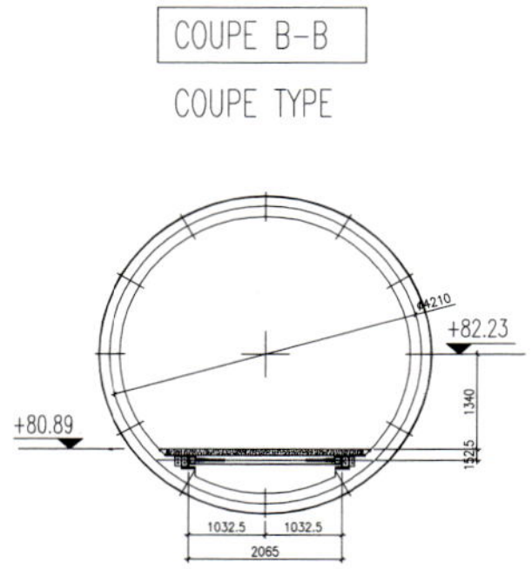

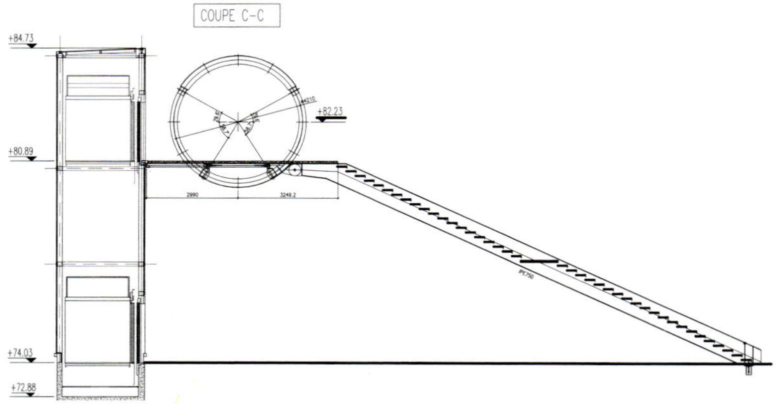

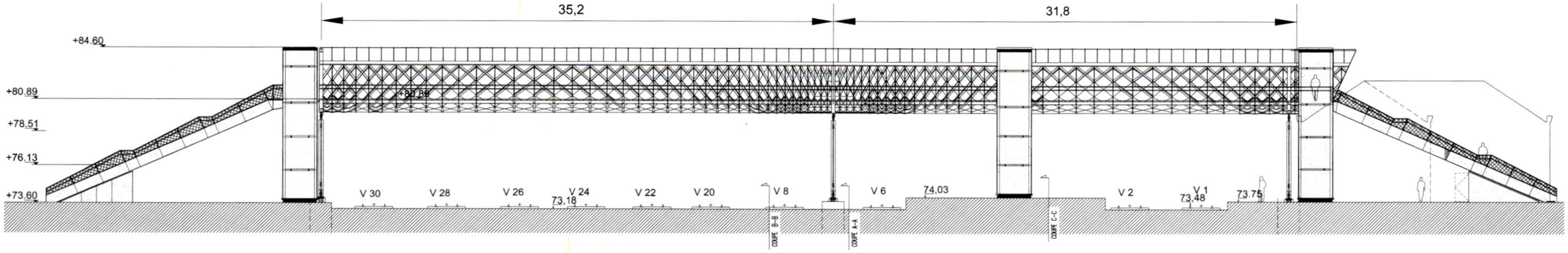

立面图 /Elevation Drawing

设计细节

俗话说“无运动不建筑”。人行天桥也不是一个静止的物体。不管是从使用功能看，还是从城市角度看，它都动感十足。设计师采用结构体系和立面饰材（交错的聚碳酸脂材料面层保护行人免受恶劣天气影响，同时可以随着结构韵律发光。）来表达这种动感，行人走在天桥上仿佛瞬间穿越了历史与现实之间的时空隧道。鲜艳的橘红色的选用是为了强调人行天桥在城市规划中的重要性。天桥的重新设计采用统一的对角网，但整体呈管状式，形成一个完整的圆柱体，供路人穿行。因人行天桥位于铁轨之上，要求其建筑设计必须做好侧面防护以确保行人和桥下火车的安全通行。本项目采用单一的结构方式和必要的惯性横跨，为防护屏障和遮篷提供支点。

Nebuta-no-ie Warasse

睡魔节博物馆

Project Location: Aomori Prefecture, Japan
Architectural Design: Molo Design
Project Area: 13,011 sq. m.
Completion Year: 2010
Client: City of Aomori, Japan
Building materials: Steel

项目地点：日本青森市
建筑设计：Molo Design 建筑设计
项目面积：13,011 平方米
竣工时间：2010 年
开 发 商：日本青森市
建筑材料：钢材

The Combination of Color and Shape 色彩与形状的衬托之美

Project Description

In March 2009 construction of this unique and sculptural building began on its waterfront site in Aomori City. Twelve-meter-tall vertical ribbons of steel encircle the building. At points they are swept aside at ground level to create openings for people to pass through. The ribbons are also individually twisted and angled to emphasize particular views and reflect colored light into the building.

Design Details

Each ribbon is given subtle variation and each catches light in a slightly different way, creating differences in color and a shifting pattern of light and shadow those changes through the day. Although there is variation between the ribbons, each ribbon is shaped to respond to the ribbons on either side, creating a varied yet unified, harmonious effect overall.

east elevation

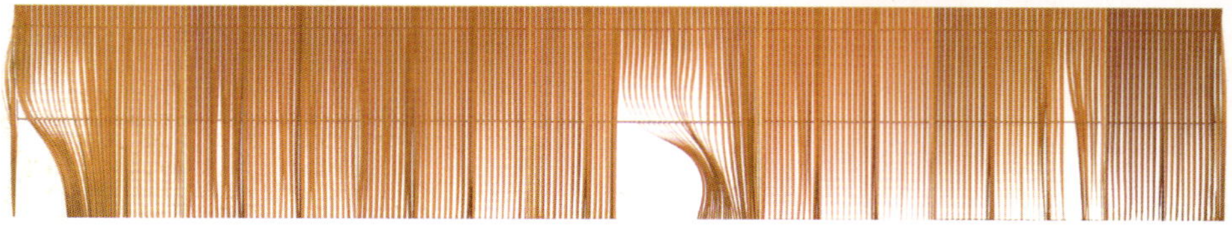

south elevation

项目介绍

睡魔节博物馆位于青森市的火车站前方，靠近海边，于 2009 年 3 月对公众开放。在博物馆的广场上可观赏到青森湾的景色，一面用钢条制作的 12 米高的幕墙包裹着整座建筑和一条户外人行道。这也象征着睡魔世界与现代城市之间的一道门槛。每一根钢条都进行扭转和弯曲，形成引入光照和视野的开口。

设计细节

每条色带都有着微妙的变化，以吸收不同角度的阳光，反射出一天中不同色彩的阳光和阴影。虽然每条色带之间都有着微妙的变化，但每一条的形状都与两边的色带相呼应，创造一个丰富多彩且独一无二的和谐效果。

ミュージアム

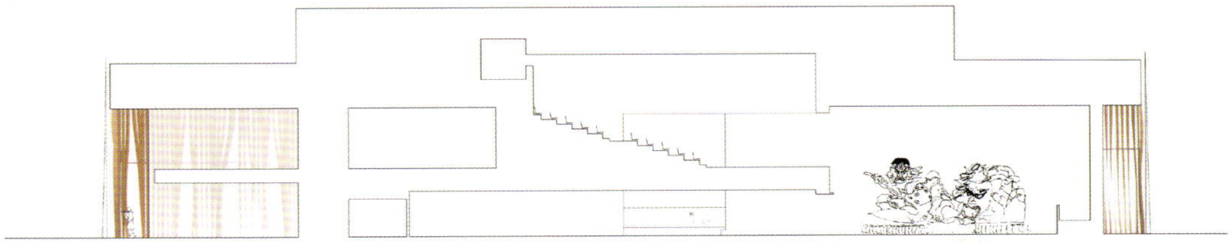

section c

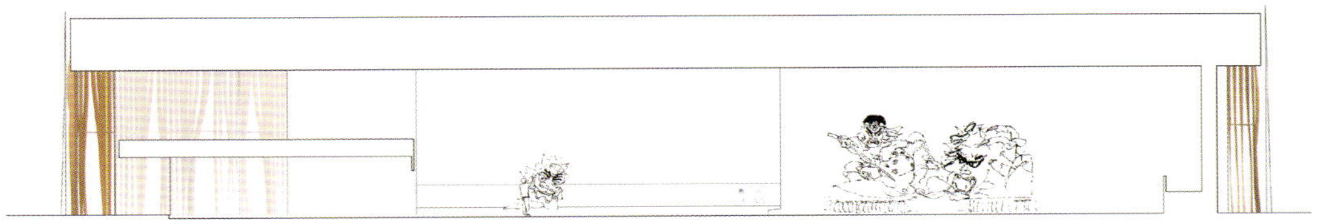

section d

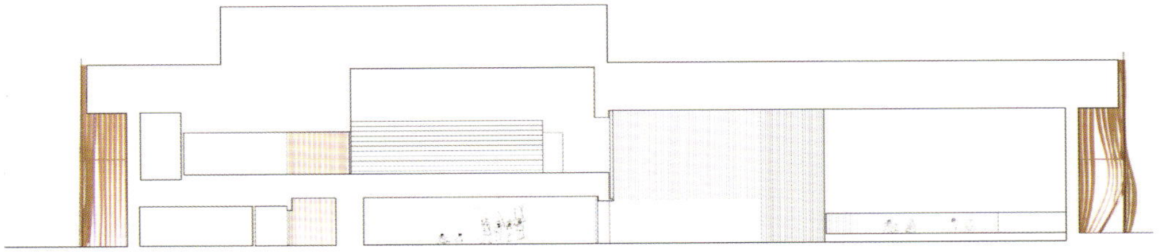

section a

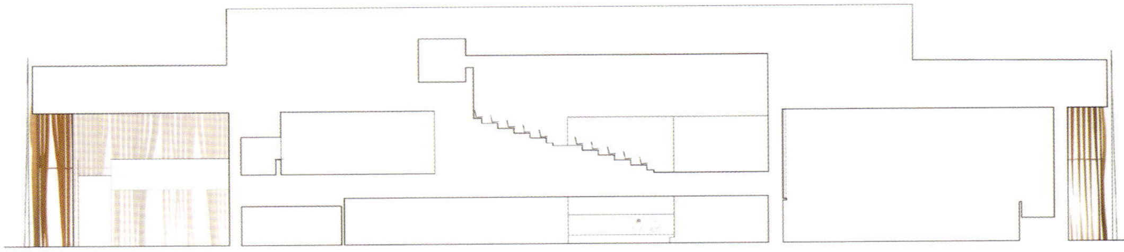

section b

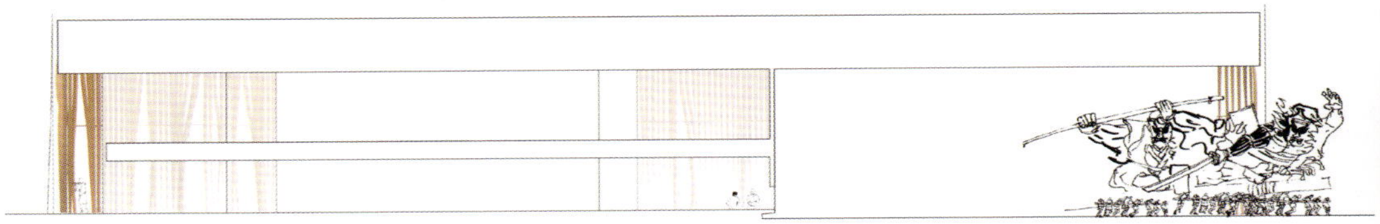

section e

剖面图 /Sectional Drawing

Ski Jump

滑雪跳台

Project Location: Oslo, Norway
Architectural Design: JDS Architects
Project Area: 30,000 sq. m.
Completion Year: 2011
Client: Oslo Municipality
Photography: JDS Architects

项目地点：挪威奥斯陆
建筑设计：JDS Architects 建筑事务所
项目面积：30,000 平方米
竣工时间：2011 年
开 发 商：奥斯陆市政府
摄　　影：JDS Architects 建筑事务所

Oslo Entertainment Space 奥斯陆最新娱乐场所

Project Description

Looking at the Olympic models of Barcelona, Sydney and Munich, architecture was at the forefront of redefining how the city was perceived. Designers believe that the Holmenkollen ski jump could liberate Oslo from regional perceptions and help leverage a more international status. The ski jump also invigorates the capital with an icon, adding to the city's collection of exemplary civic buildings like the new opera house and reinforcing the city's wave of regeneration. There is another crucial aspect to this iconographic issue that the project emphasizes: from its strategic position, at the highest peak over the city, the structure offers the most breathtaking views of Oslo, the fjord and the region beyond. They have designed a public square, on top of the jump, literally a plaza, to let visitors experience Oslo from this exceptional vantage point. It's a new form of public space, using an unlikely architectural form as its host—allowing for a dialogue between the city and its inhabitants.

Design Details

The design aims at unifying the various elements present in a ski jump into one single expression of shape and action. Rather than having a series of dispersed pavilions on site, they've managed to combine them into one organism. The judges booths, the commentators, the trainers, the royal family, the VIPs, the wind screens, the circulations, the lobby, the entrance to the arena and the arena itself, the lounge for the skiers, the shop of souvenirs, the access to the existing museum, the viewing public square at the very top, everything, is contained into the shape of the jump in a symbiosis of programs and experiences, giving the jump a harmonic contextual relationship to the surrounding landscape. The resulting simplicity of the solution improves the experience of the spectators and the focus of the skiers. There's an impressive feeling of intimacy at Holmenkollen: both audience and jumpers are enclosed in the arms of the ski jump in a form of an architectural embrace.

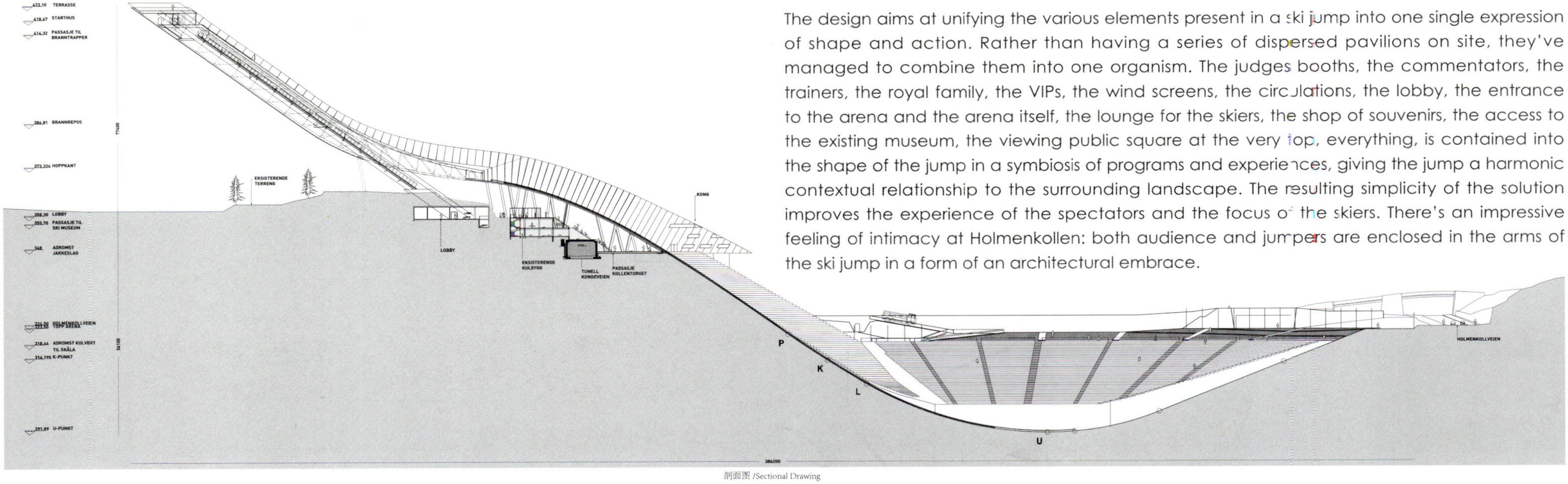

剖面图 /Sectional Drawing

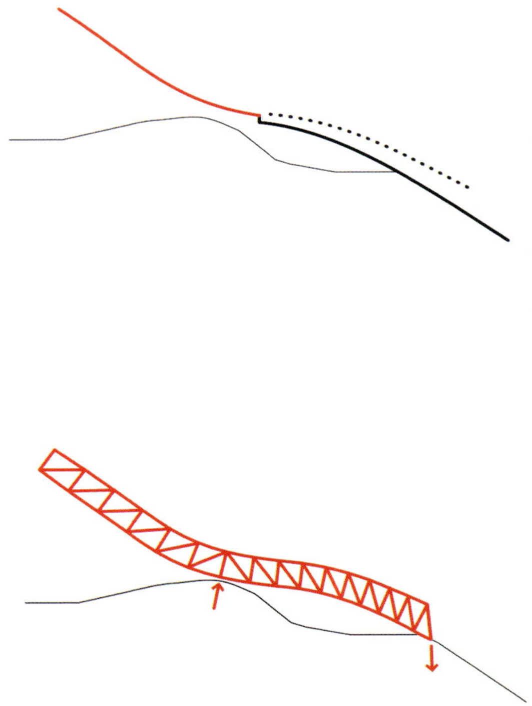

项目介绍

纵览巴塞罗那、悉尼和慕尼黑的奥运模式，建筑设计的首要目的是重新定位人们对城市的感知度。设计师们坚信霍尔门科伦山的滑雪跳台可以将奥斯陆从局部感知解放出来，有助于更好地充分利用其国际地位。滑雪跳台作为城市的新标识也为首都增添了活力，如新歌剧院为城市增加了模范建筑群，加强城市的再生浪潮。关于肖像权方面，还有另外一个重要的问题是该项目所强调的：从它的战略地位来看，它位于城市的最高峰，该项目的架构为人们提供奥斯陆、海湾及该地区最秀丽的风光。设计师设计了一个名副其实的广场，在跳台顶端，以便用一种不太可能的建筑形式做主体，让城市与它的居民之间进行对话。

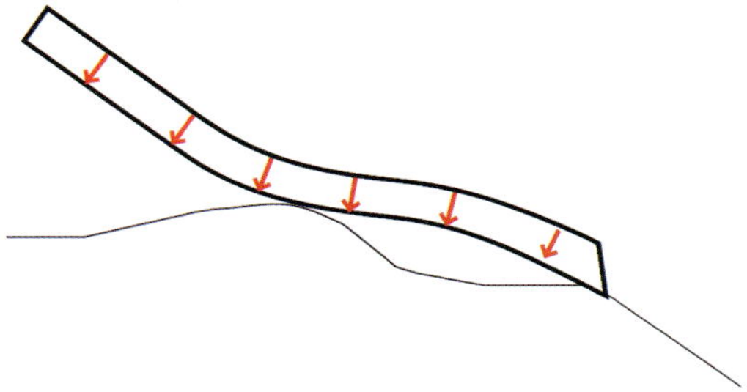

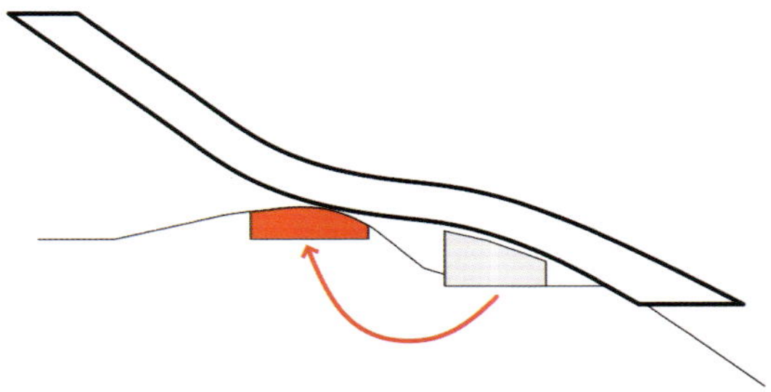

设计细节

该项目的设计旨在于一个跳板上统一不同的元素，使之成为一个简单的外形和设计的表达，而不是就地设计一系列分散的亭子，设计师设法将它们合并成一个有机体。法官亭、解说室、教员室、皇室家族、贵宾室、遮风屏、循环设计、大堂、竞技场的入口和竞技场、滑雪者的休息室、纪念品商店、到达现有博物馆的小径，以及顶端的观众席等这一系列设计，都结合方案和经验，承载着跳跃的形式。由此产生的简单设计使观众对滑雪者的关注更加集中，令人们对霍尔门科伦山有一种无与伦比的亲切感，那就是设计师以一种建筑形式将观众和滑雪者包围在滑雪跳台之中。

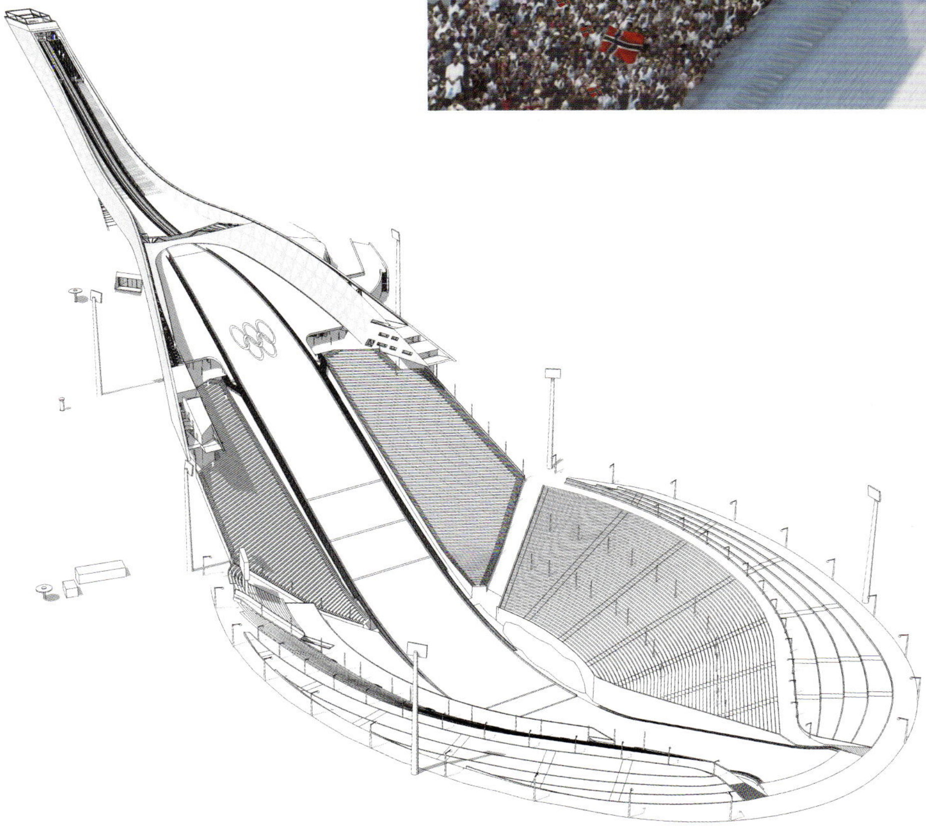

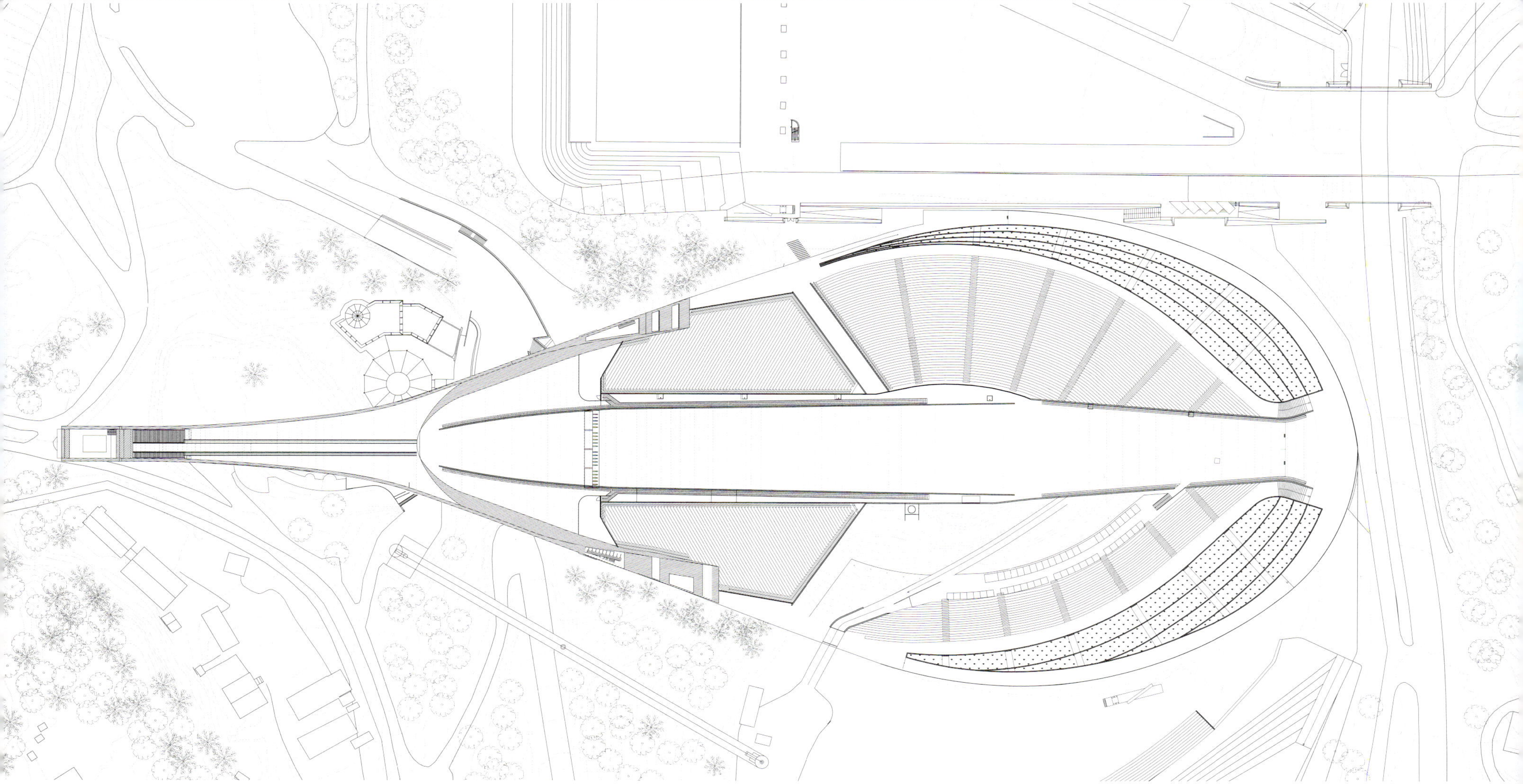

全景平面图 /Panoramic View Plan

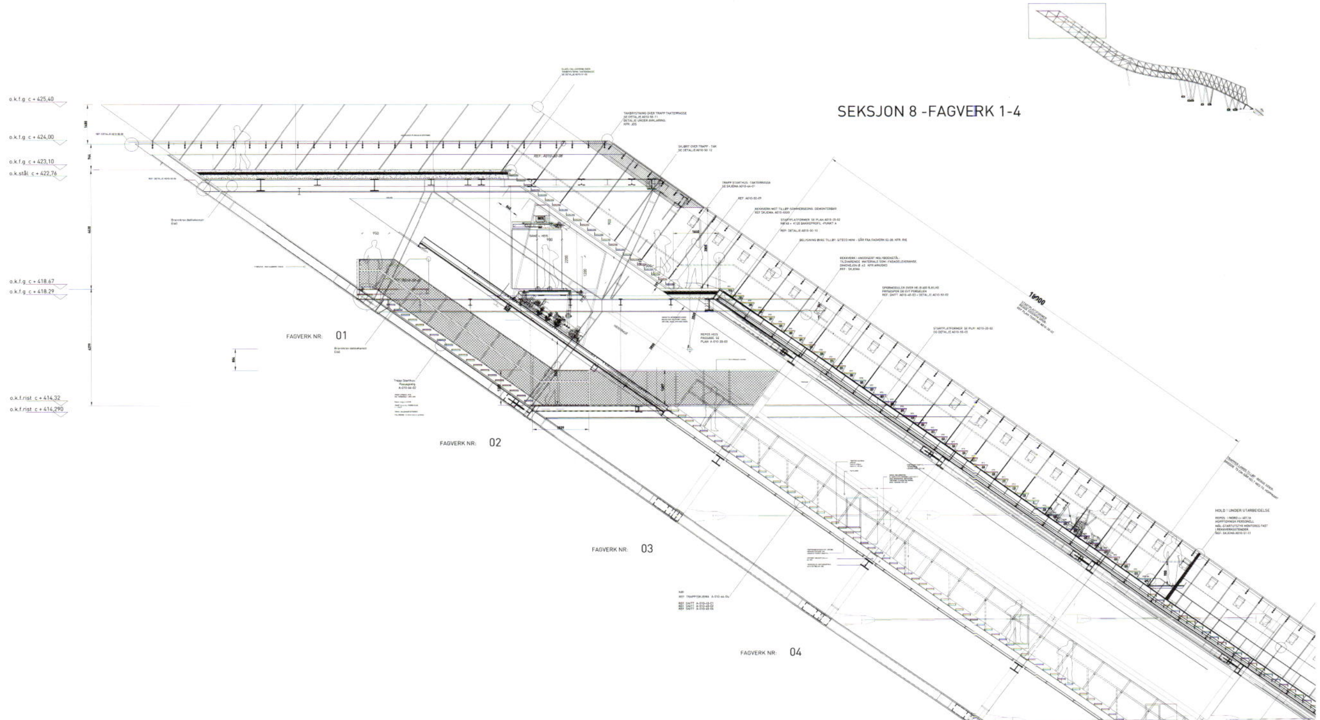

A010-41-01 LANGSNITT ØVRE TILLØP

Erasmusgracht Bridge

Erasmusgracht 桥

Project Location: Amsterdam, Holland
Architectural Design: DP6 Architecture Studio
Completion Year: 2010
Client: Gemeente Amsterdam
Photography: Daria Scagliola& Stijn Brakkee
Building materials: Wood, Steel Plate

项目地点：荷兰阿姆斯特丹
建筑设计：DP6 Architecture Studio 设计事务所
竣工时间：2010 年
开 发 商：阿姆斯特丹市
摄　　影：Daria Scagliola & Stijn Brakkee
建筑材料：木材、钢板

A Public Building in Amsterdam 阿姆斯特丹公共建筑

Project Description

The design for the bridge spanning the Erasmusgracht was driven by striking shapes and the intention to create an identity for its surroundings; shaping and revitalizing the Amsterdam neighborhoods the Kolenkitbuurt and the Laan van Spartaan. The specific contours were the direct result of the translation of its functional requirements and the forces at play. A district-defining waterway through Amsterdam, the Erasmusgracht is flanked by green slopes and umbrageous trees.

Design Details

Playing on the forces that emerge from a structure with one asymmetrical support and considering the minimum span and headroom resulted in a highly recognizable composition. The bridge consists of a lightweight construction and two primary beams at each side of the bridge deck. The beams follow the lines of momentum and translate into the bridge's handrails. Enclosed by welded steel plates with a light gray coating, -he bridge mirrors the shimmering water below and reflects varying images and plays of light off the different planes. The bridge is meant to portray a tightly crease sheet of paper and finds its lightness from these technical and material choices.

项目介绍

桥梁的设计横跨 Erasmusgracht，与周围的环境相比，独树一帜。它为阿姆斯特丹的“邻居”——Kolenkitbuurt 及 Laan van Spartaan，注入了新的活力。该项目功能上的要求及空间的限制，使得桥梁具有鲜明的轮廓。水流沿街而走，穿过阿姆斯特丹。Erasmusgracht 的侧面，山坡上种满植被和葱郁的树林。

设计细节

迫于结构上一个不对称支撑点的限制，及对最低跨度及净空高度的考虑，设计师煞费苦心，最终做出了令人满意的作品。桥梁包括一个轻量结构及桥面两端的两个主梁。焊接的浅灰色涂层钢铁板材使得桥梁映照着河水，闪闪发光，变化多端。通过技术处理和用材选择，桥梁就像是一张褶皱的纸，有很多个折射亮点。

Iglo Greve

Iglo Greve 文化运动馆

Project Location: Hundige, Denmark
Architectural Design: CEBRA
Completion Year: 2008
Project Area: 1,650 sq.m.
Photography: CEBRA

项目地点：丹麦
建筑设计：CEBRA 建筑事务所
竣工时间：2008 年
项目面积：1,650 平方米
摄　　影：塞夫拉建筑事务所

A Unique Sports Center 一个独特的体育中心

Project Description

In 2008 the municipality of Greve commissioned CEBRA to draw the Iglo at Hundige Boldaelled. The design was developed through a dialog-based development process including users and stakeholders. At the end of the design phase the project was taken over by a construction team with Nova5 as architects.

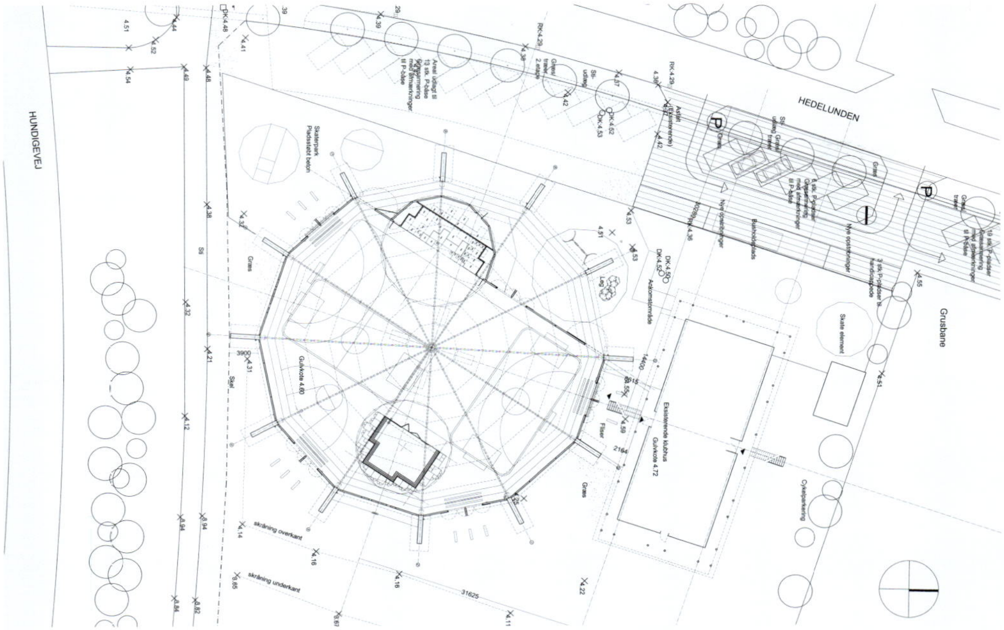

全景平面图 /Panoramic View Plan

Design Details

A sports hall like this can end up anywhere and is never shaped by its surroundings. The round dome shape has no direction and its strong geometry makes it independent and harmonious, like an igloo. The concept for the Iglo was developed in 2003 by CEBRA in collaboration with The Danish Foundation for Culture and Sports Facilities, as an example of a cost-efficient, un-heated sports facility. The building works in principle as a covered out-door facility that isn't heated but protects from rain, wind and snow. The facility can be used for multiple sports—and cultural functions.

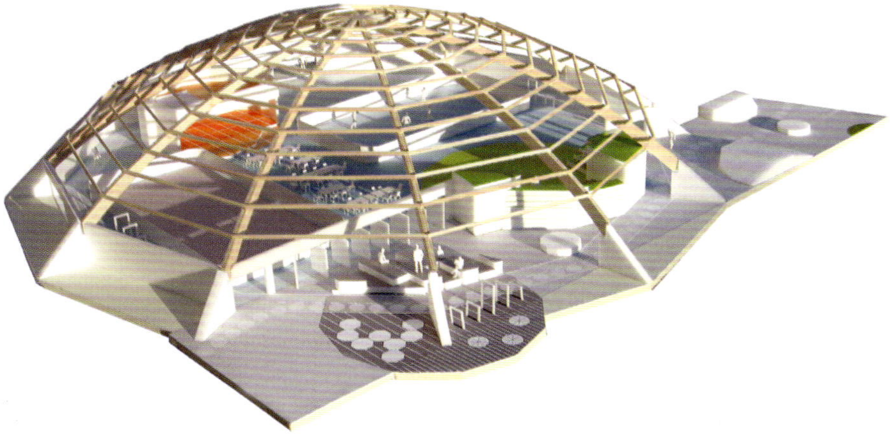

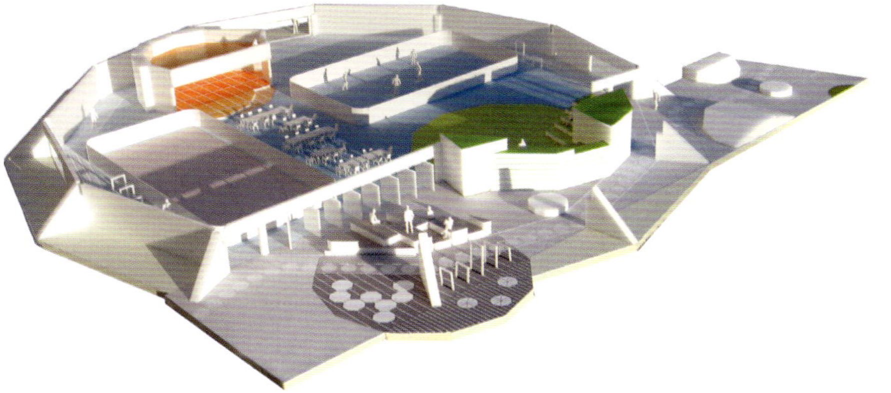

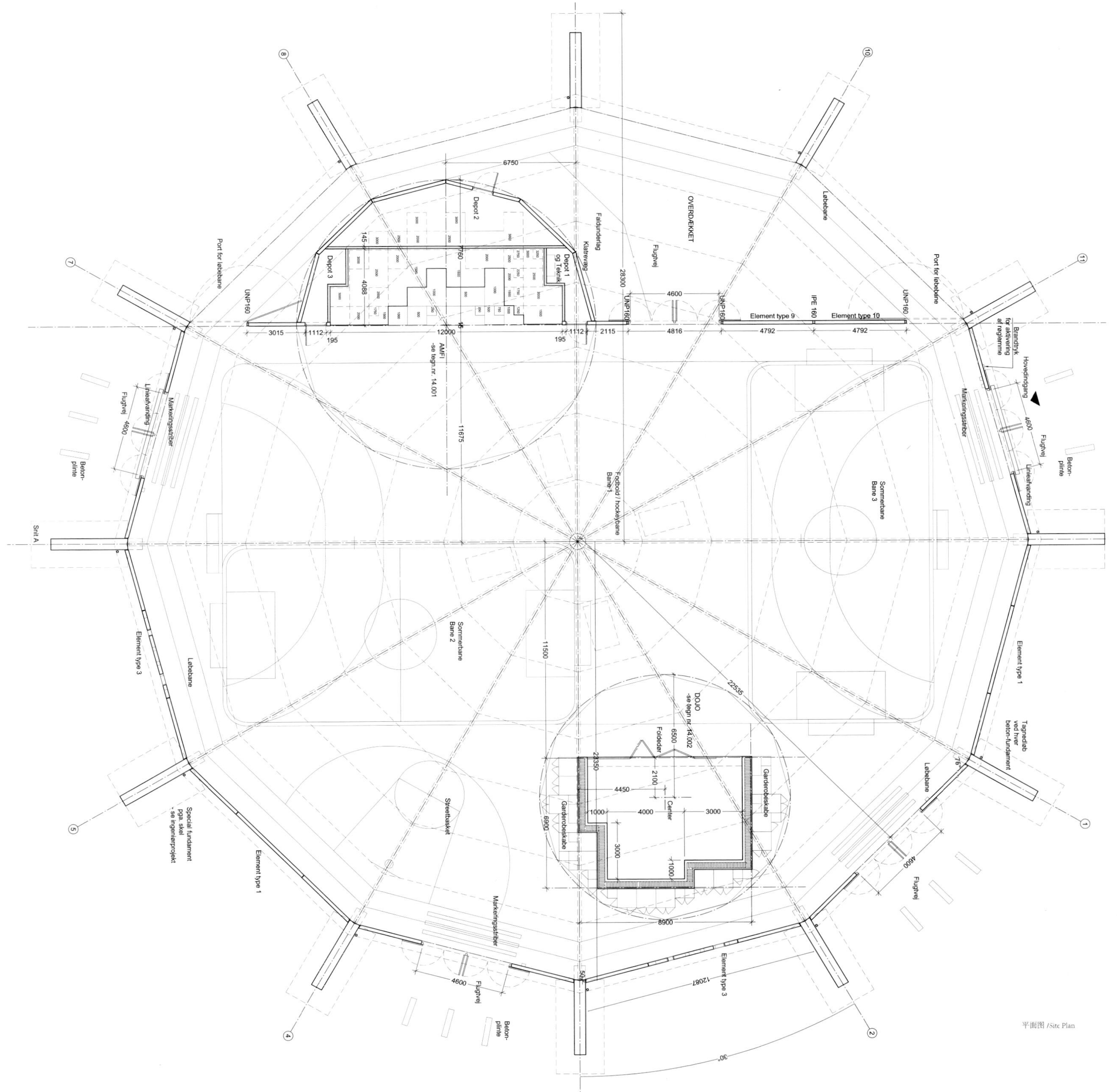

平面图 /Site Plan

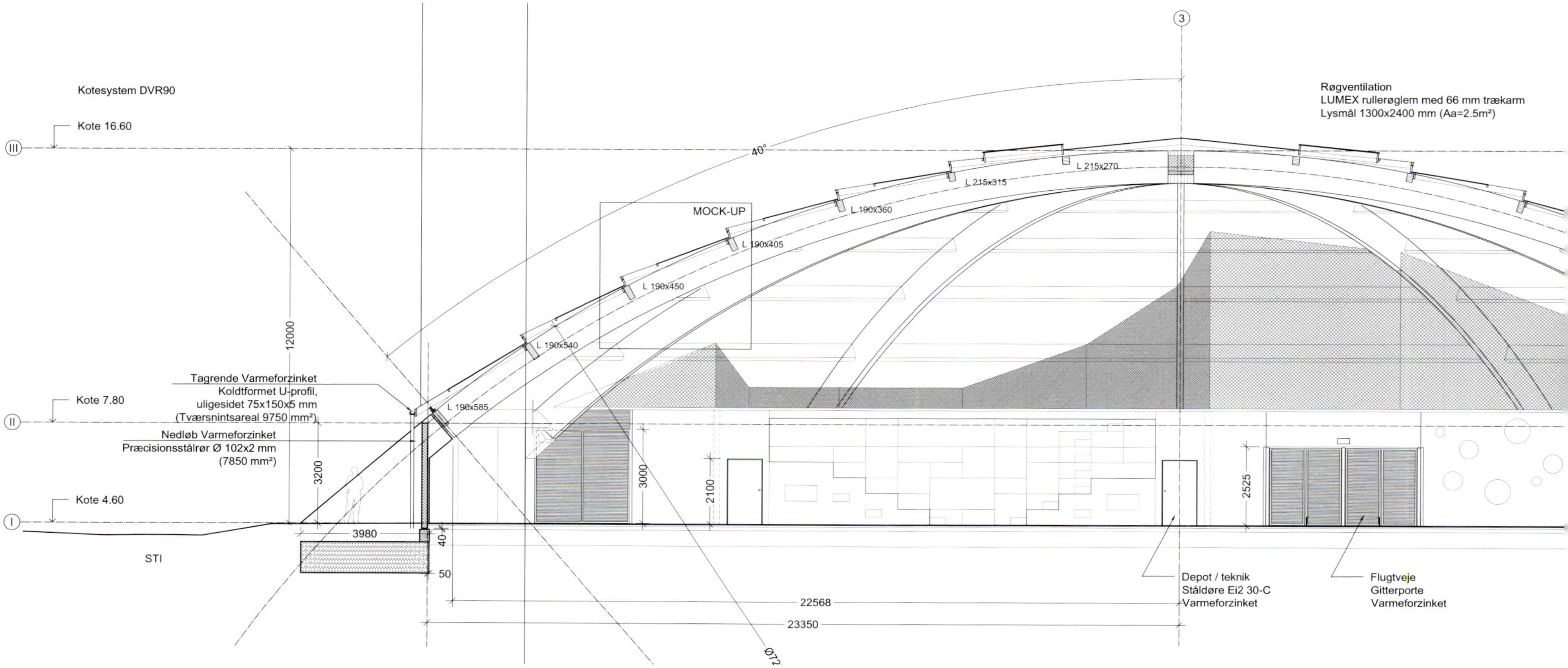

剖面图 /Sectional Drawing

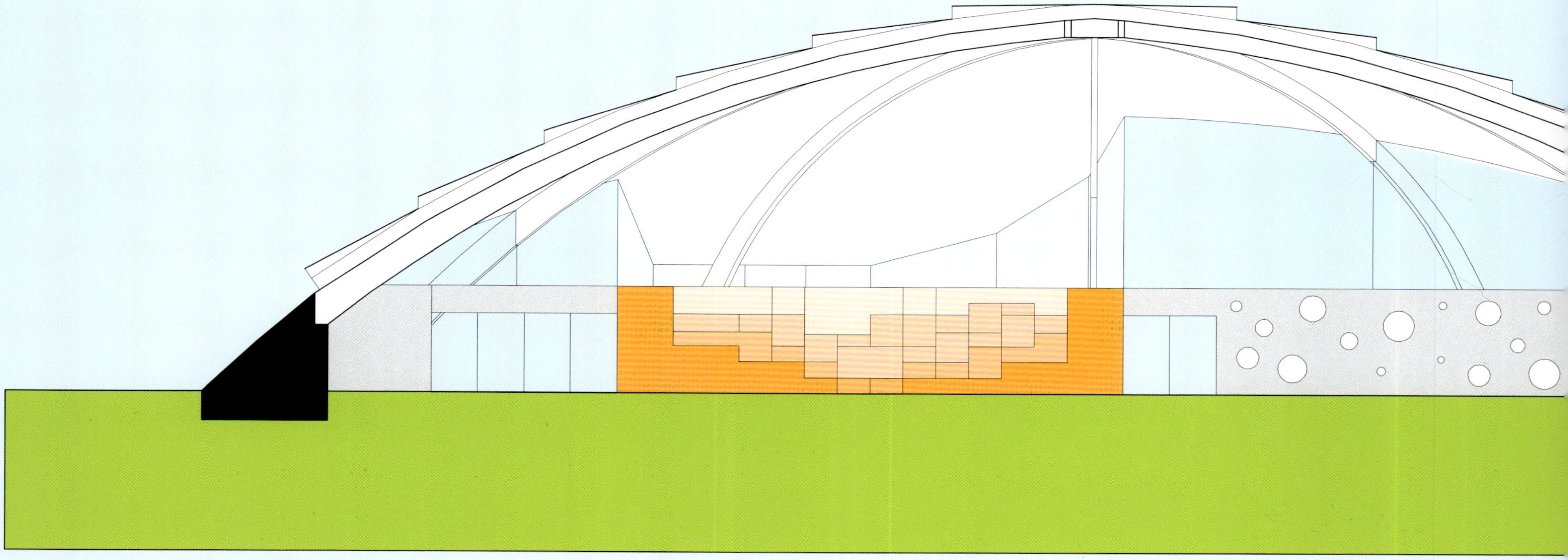

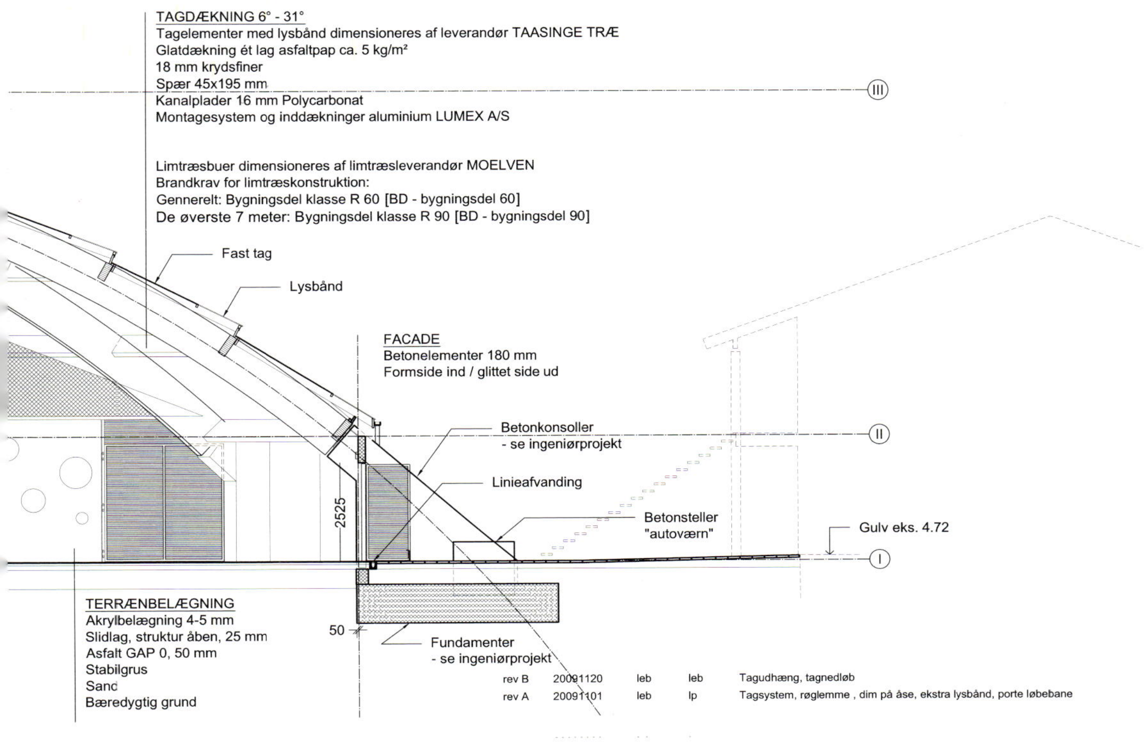

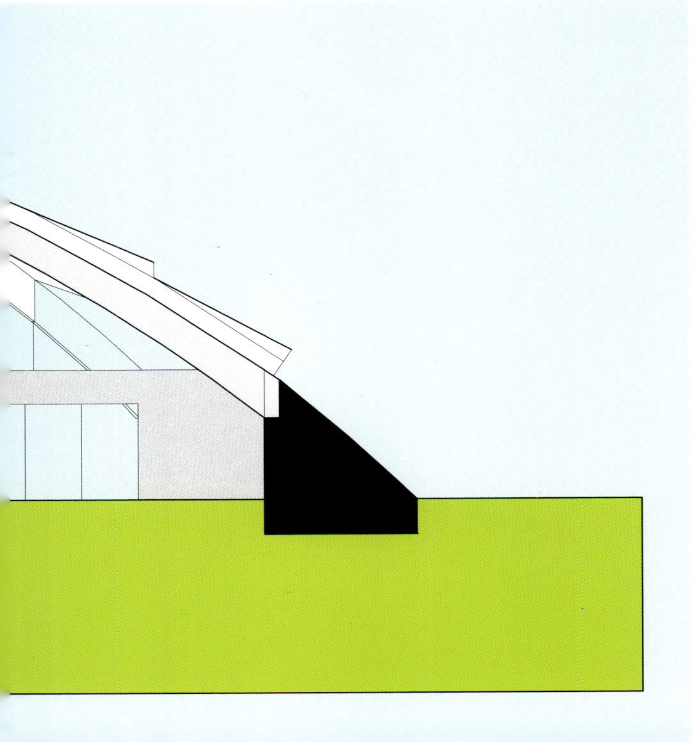

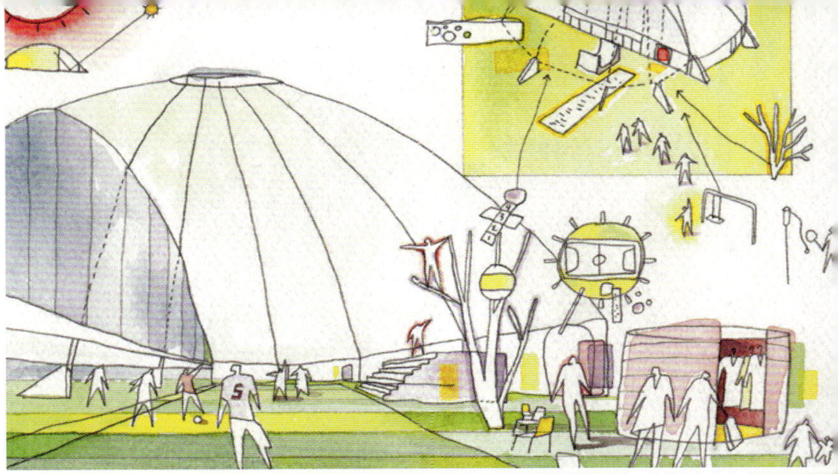

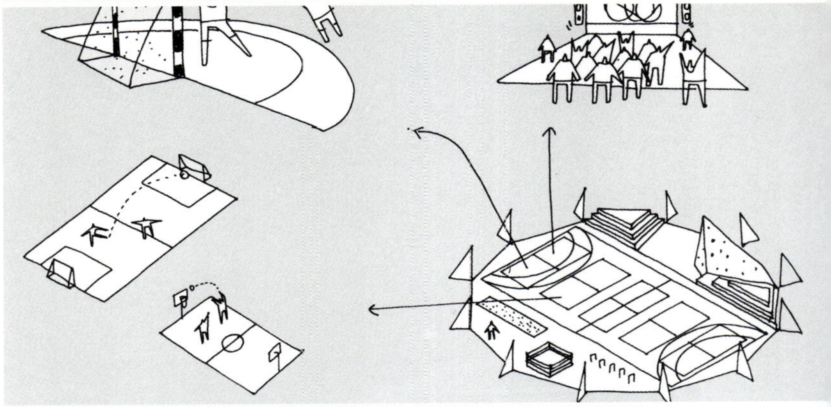

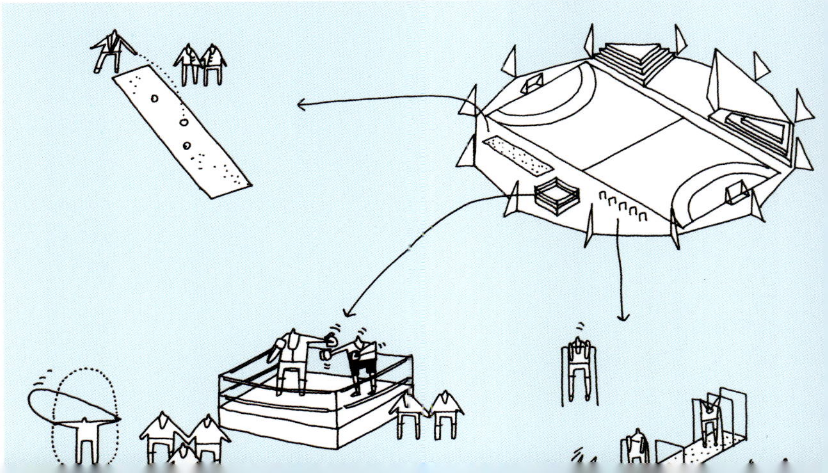

项目介绍

2008 年，丹麦格雷沃政府当局委托 CEBRA 建筑事务所为 Hundige Boldfaelled 设计一个文化运动馆。该项目的整体设计综合了用户和投资商双方的要求。在最后的设计阶段，本项目被 Nova5 建筑设计团队接手。

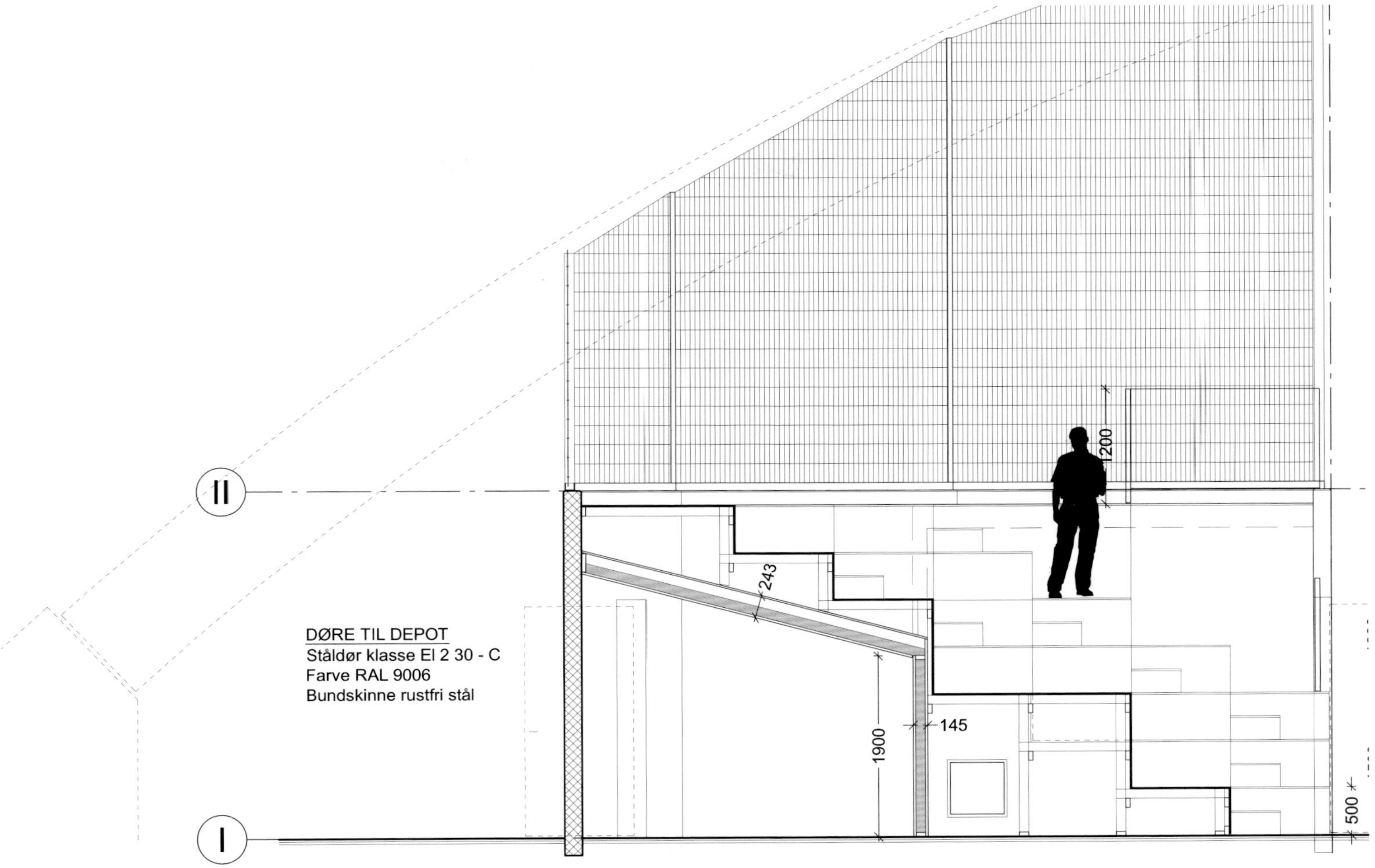

剖面图 /Sectional Drawing

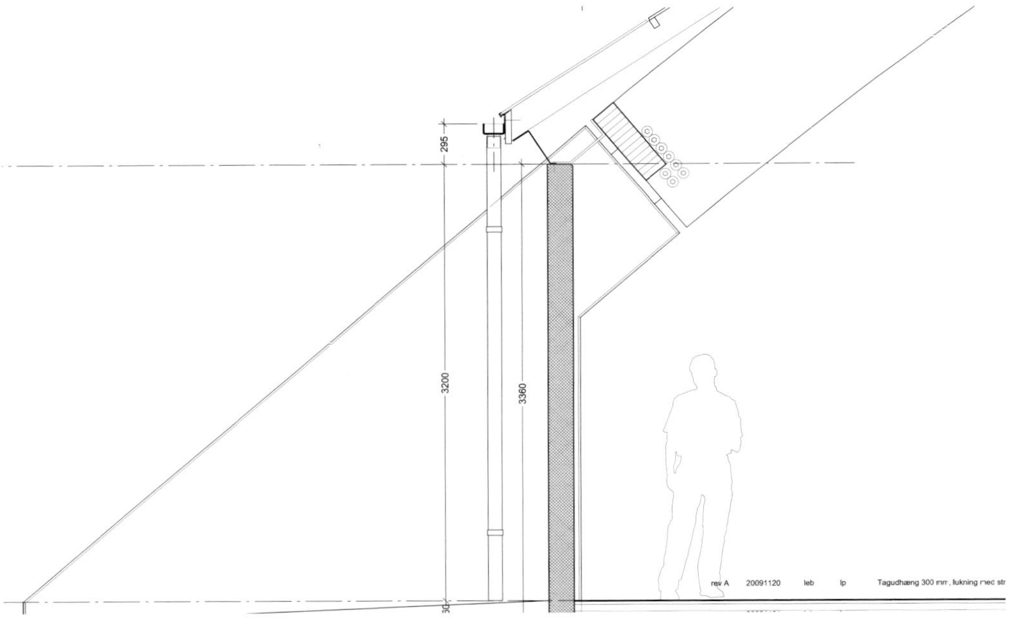

设计细节

这样的体育中心可以任意置放，外形的设计不需要考虑周围的环境。圆屋顶没有方向性，但其强烈的几何感独具个性，就像是爱斯基摩人的圆顶冰屋。这一设计理念是 2003 年由 CEBRA 建筑事务所与丹麦文化体育设备基金会共同合作完成的。这一具有成本效益的建筑虽然没有供暖设备，却可以防风、遮雨雪，也可以举办各种体育文化活动。

The Lantern Pavilion

灯笼亭

Project Location: Langgata, Sandnes, Norway
Architectural Design: AWP /Atelier Oslo
Project Area: 900 sq.m.
Completion Year: 2010
Client: Sandnes Municipality
Photographer: AWP+ATELIER OSLO
Building materials: Wood, Glass

项目地点：挪威桑内斯龙街
建筑设计：AWP/Atelier Oslo 设计公司
项目面积：900 平方米
竣工时间：2010 年
开 发 商：桑内斯政府
摄　　影：AWP/Atelier Oslo 设计公司
建筑材料：木材、玻璃

Perfect Expression By Ancient Wood With Fashion

古木建筑与时尚的完美展现

Project Description

A manifesto for public space design and an urban strategy, deal with the Norwegian ancient wood architecture motives to design an urban lantern. A wood and glass giant canopy, the "Lantern" displays an open roof of such a dimension that it becomes the symbol of this city. It also protects a new market square for which they designed pavements and facilities allowing different functions: weekly market, concerts, events, loitering and exchanging. Lit at night, the Lantern has become the City´s new icon and landmark providing an inspirational atmosphere to sustain existing practices and encourage new ones.

Design Details

The proposal used the iconographic shape of an old, uplifted wooden house. By the redefinition of its traditional construction and by making it glow in the dark, a new landmark for the city was created, a symbol of the old city upscale to the new city's dimensions, dealing with the Norwegian wood ancient architecture motives to design a contemporary object. This public Lantern aimed at providing a space where to enjoy light and weather changes if/when nothing else takes place. And if there is always something happening, on account of the Lantern, more will happen: creativity calls for creativity.

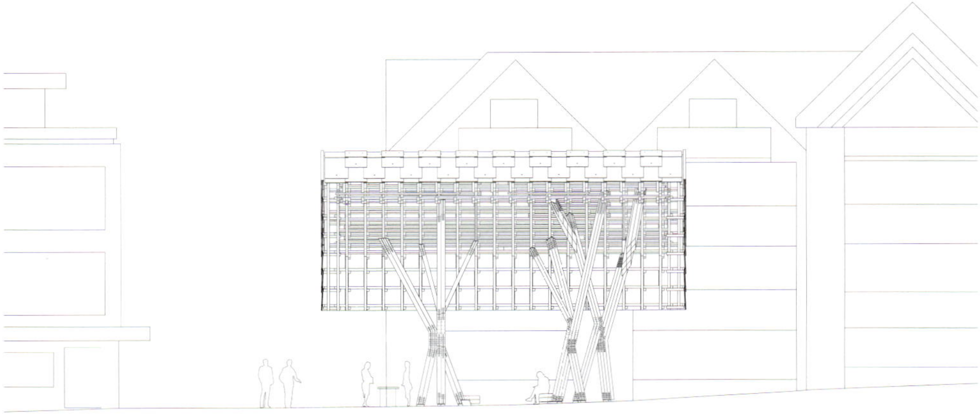

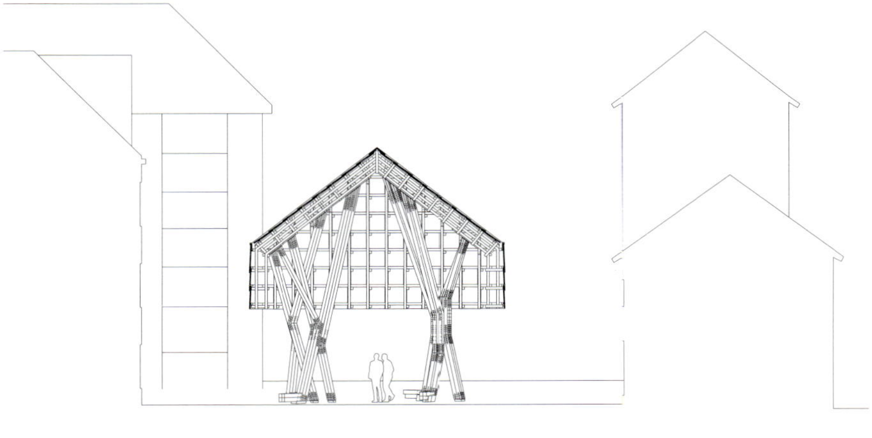

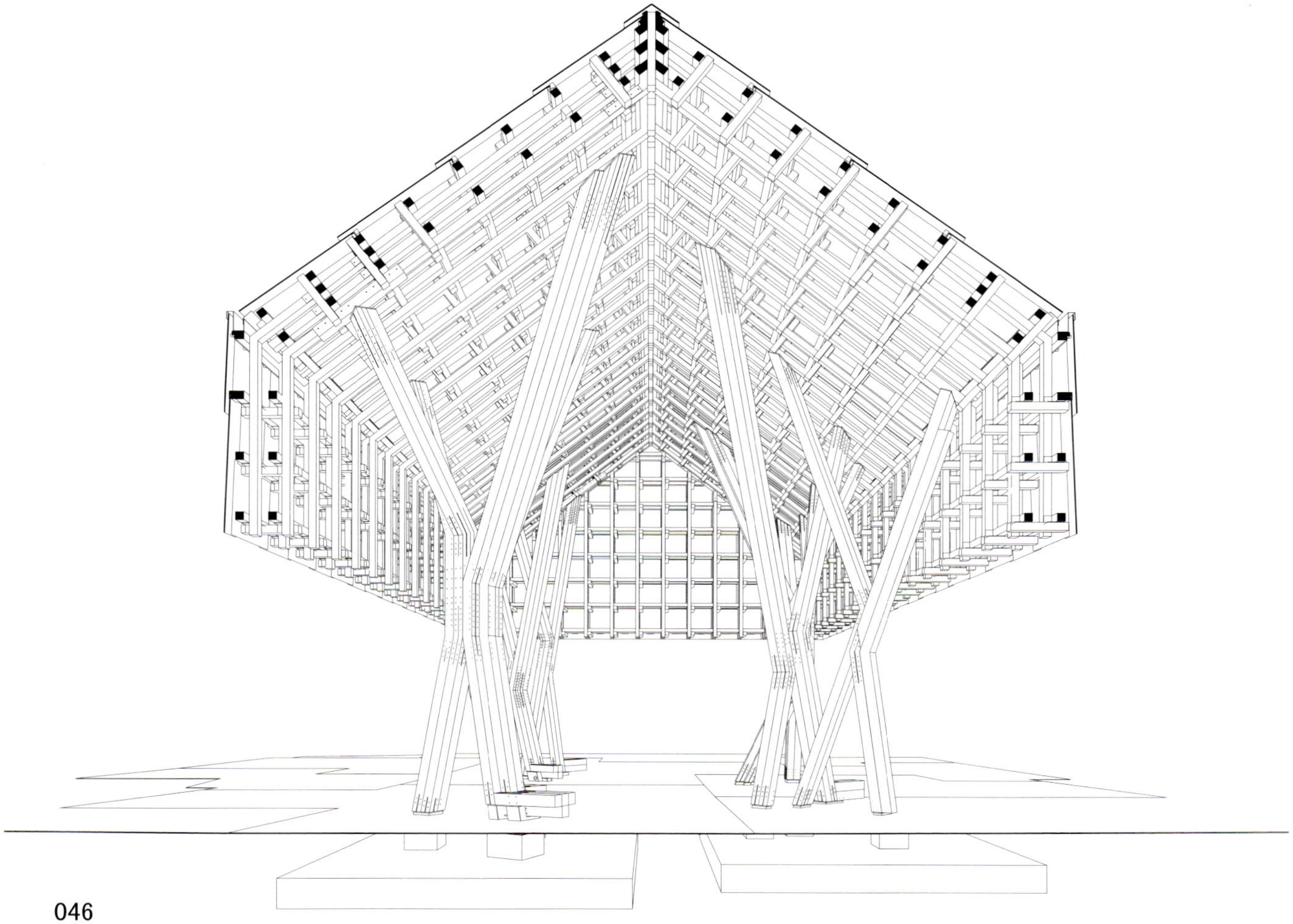

Columns create benches, and are tilted to give lateral stability.

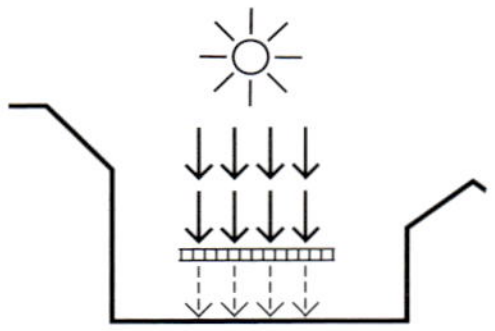

The project has a transparent structure in order to give natural light under the shelter.

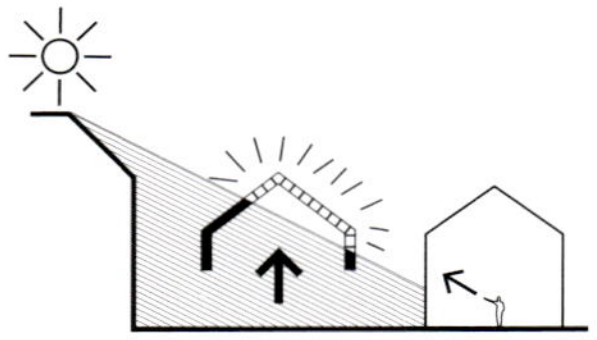

The structure has height so that it catches sunlight and creates visibility.

The translucent glass pattern and the structure creates a lantern effect at night.

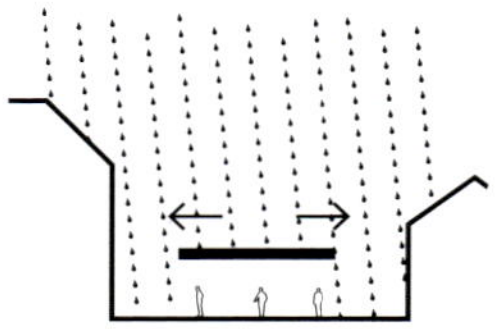

In order to get the most functional space, the project covers the largest possible area allowed.

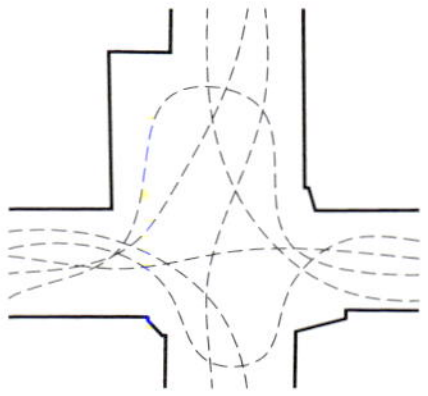

A continuous floor creates a flexible space.

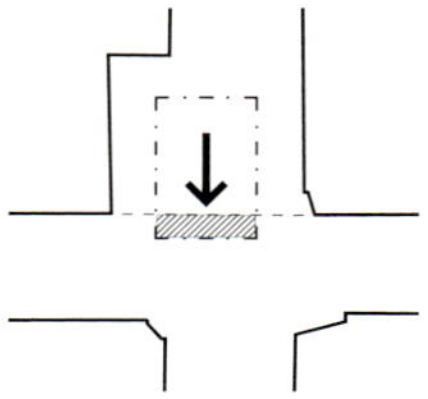

The structure is positioned to create visibility.

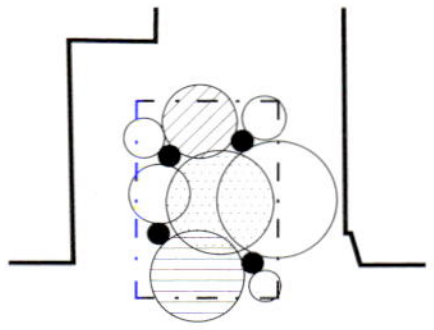

The positions of columns create a sequence of overlapping spaces.

项目介绍

该项目的设计宣言和规划意图是结合挪威的古木建筑，为桑内斯市设计一个公共建筑。该项目的设计结合了两个因素，即在雕塑般的木架结构上，建造了一个时尚的玻璃屋顶，来映照城市清晰的面貌。设计师将屋顶用四组橡木支撑，并在角落里摆放了长凳，供居民和游客休息。城市的改变，就从一个发光的角落开始。这座位于挪威桑内斯市的木建筑，名为灯笼亭，由国际知名的 Atelier Oslo 与 AWP 两家建筑事务所共同合作规划，借其创新来展现挪威的古木建筑所蕴含的传统精神，重塑该市的街区形象，同时也呼应了桑内斯曾获得“欧洲文化之都”的名号。

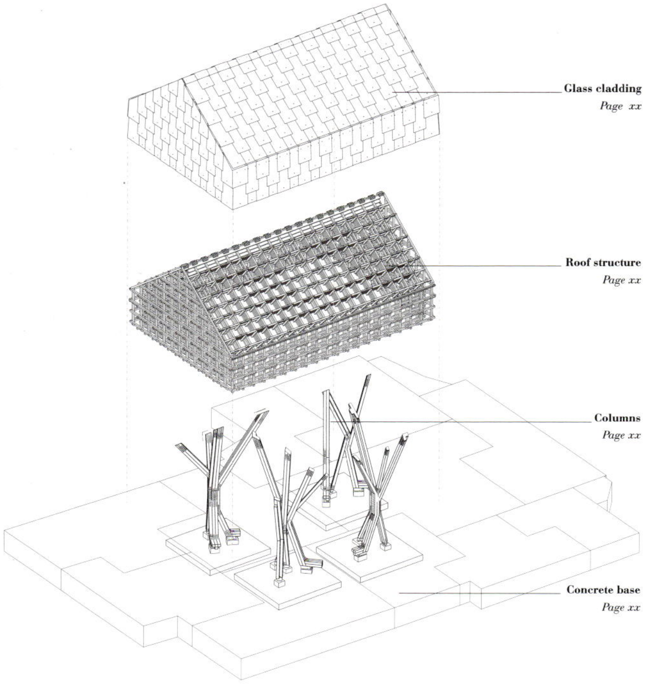

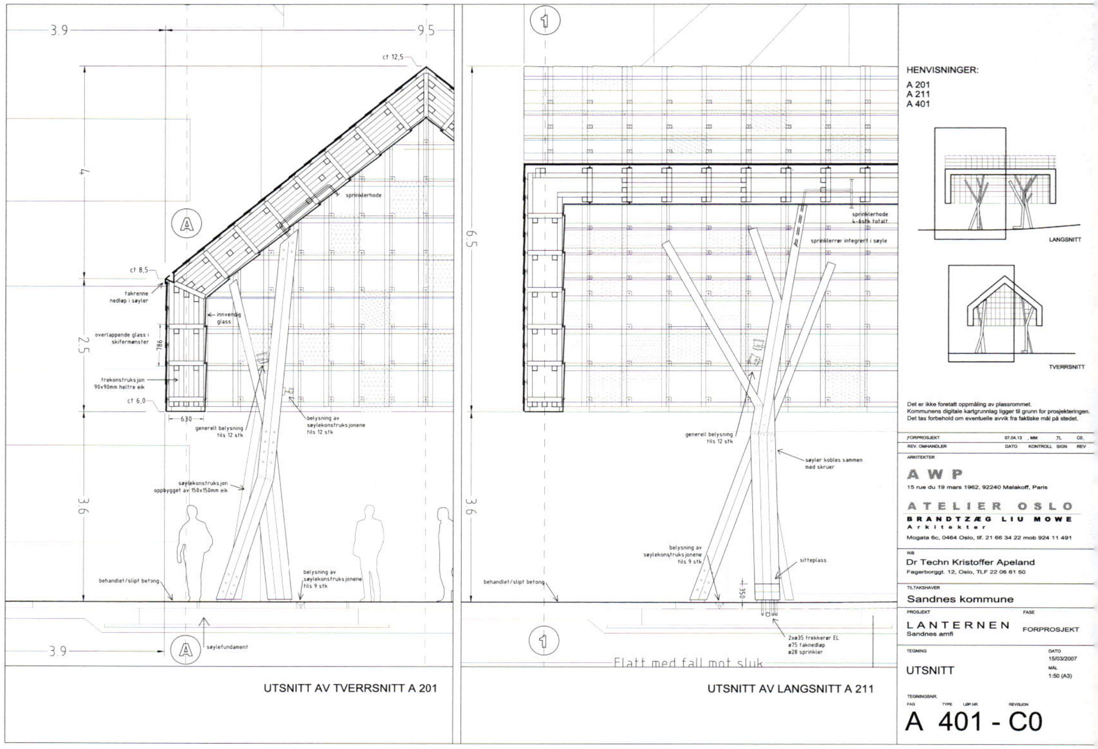

设计细节

灯笼亭看似长了脚的房子，又像悬浮在半空中的老式木屋，其实是建筑团队采用挪威传统木屋的形象，以艺术线条与大量玻璃材质演绎的经典，并已成为街区的标志性建筑。在建筑团队的创意构思下，他们用交叠的鳞片状玻璃，取代传统屋顶上的瓦片，使视觉通敞无碍，更让喜爱阳光的北欧人在此休憩时，依然可以享受阳光的照耀；若是阴雨天气，也无损于它遮风避雨的功能。灯笼亭的双层式格栅结构，充满韵律的规则美感，架构出质轻、抽象的理念，与木材的温润感共同产生和谐的氛围。它更在筛落阳光时，在磨砂地面幻化出多变的影像。灯笼亭灯光的角度设计与架高屋体的柱脚，让它在夜晚呈现华丽样貌。

OTTEREN

Nautilus

贝壳馆

Project Location: Naucalpan, Estado de Mexico
Architectural Design: Javier Senosiain
Project Area: 330 sq.m.
Completion Year: 2007
Photography: Jaime Jacott
Building materials: Ferrocement

项目地点：墨西哥
建筑设计：贾维尔•赛诺西设计公司
项目面积：330 平方米
竣工时间：2007 年
摄　　影：Jaime Jacott
建筑材料：钢丝网水泥

A New Trend in Residence 住宅的另一走向

Project Description

At the end of the turn-around is the piece of land, with upward topography, where the Nautilus was built. It is limited by three of its adjoining properties because each of them has high buildings. The fourth adjoining property if to the west and has wide views of a green area with mountains in the horizon. The construction area was defined since the first studies at the back of the piece of land, leaving the pedestrian and car access at the front as well as only one facade, the so called fifth facade in architectonic language.

To develop the design, shaping the area that will be inhabited designers start from a deep observation of Nature's principles. An observation will let them to perform, as Bachelard would say, a poem of space where the concept flows from the inside, an organism dwelling. That is how it arouses, almost spontaneously, nothing is imposed. They find the shape instead of giving it shape, it is released from the outside; in other words, they attempt to help things to discover their shape. Here the ground is spiral. The spiral appeared, it adapted to the land and for outline they draw such spiral, logarithmic. The model's work led them to different changes until they realized that the volume asked for the Nautilus.

Design Details

The grounding, walls, floors, ceilings and furniture work like a 4 to 5 cm thick ferrocement eggshell, what gives the structural rigidity is the main shape in continuous spiral, of double curvature. Literally, the dwelling place is a big snail structurally working as such, an evocative shell: resistance, protection, shelter...For ventilation, there are two underground ducts, when the outside air enters, the house gets cold or hot according to the year's season, during hot weather the air is chilled and refreshes the house flowing through the snail in spiral; pushing the hot air upwards and going out through the high part of the Nautilus. On the other hand, when the outdoor temperature is cold, the air that comes in through the duct warms the house.

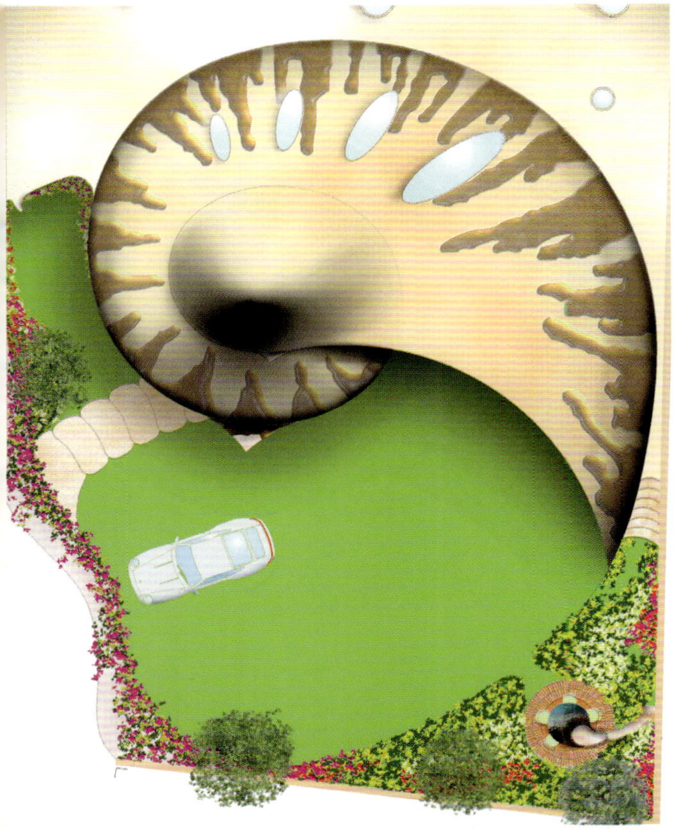

项目介绍

贝壳馆的地理位置呈向上的斜坡式，由于周边建筑均为高层建筑，因此它的架构受到三栋毗邻建筑的限制。作为第四栋加入该地区的建筑，如果矗立于西面，将拥有宽阔的视野来享受苍翠碧绿的山脉。第一次在研究这片土地的背景情况时便确定了它的位置。设计师决定将前面的人行道和车道作为单独的一面，在建筑结构上被称作是第五面。

为了实现它的外形设计，设计师要从自然原则开始深入观察，并将观察结果表现出来，正如巴舍拉尔会说的，诗意化建筑的设计理念从内向外涌出，即一种有机体住宅。那便是设计的奥妙所在，几乎同时，没有什么东西是强加上去的。设计师发现了建筑的结构，而不是主观地赋予它一种结构。换个方式说，设计师尝试着帮助事物发现其自身形状。地面呈盘旋状，盘旋状的土现使它适应该地理位置。设计师列出了盘旋状外形并不断修改模型，直到他们意识到鹦鹉螺状式设计符合其地形。

设计细节

接地装置、墙面、地板和家具就像 4~5 厘米厚的钢丝网水泥蛋壳，它给予了双曲率结构刚度。照字面上来说，该住宅的位置就是一个鹦鹉螺式结构，令人回味无穷的外壳：既可以抵抗外界，又可以保护自己的遮蔽场所。在通风方面，设计师采用了两条地下管道，当室外空气流通进来，室内便根据天气的变化变得凉快或者暖和。炎炎夏日，空气变得凉爽清新，在鹦鹉螺状式房间内流动，并且将热空气推向上方，从鹦鹉螺的顶部流通出去。另一方面，当室外温度变冷，空气通过管道流通来温暖整间住宅。

The Mushroom

蘑菇馆

Project Location: Mexico
Architectural Design: Javier Senosiain
Completion Year: 2008
Building materials: Ferro-Cement

项目地点：墨西哥
建筑设计：贾维尔 • 赛诺西设计公司
竣工时间：2008 年
建筑材料：钢丝网水泥

A Central Building in Soccer Club 足球俱乐部的中心建筑

Project Description

This soccer club, consisting of three fields, is located in Acapulco's Diamond Area. The building rises in the field's center. An oval delimits the services: dressing rooms, bathrooms and administrative offices, protected by a slope of grass acting as insulator of the region's high temperatures.

Design Details

A monolithic structure rises from this building rises, from the bottom of the land and from the platform, similar to an eruption that outcrops in the space drawing, as the impressionist did, a mushroom that in the sunlight melts its contour due to its own iridescence. Because it is a construction in ferro-cement, the mushroom can be anchored to the land. From which, a stair rises at the center, a vertical element that communicates the dressing rooms with the platform and the cafeteria, which seems to float like a flying saucer, perceiving from that point a panoramic view that allows watching the soccer matches.

项目介绍

该足球俱乐部包括三块场地，坐落于墨西哥南部的阿卡普尔科钻石区。该建筑矗立于足球俱乐部中心，用一个椭圆作为服务区——更衣室、洗手间及行政办公室。种满青草的斜坡，在一定程度上抵御了当地的高温。

设计细节

庞大的结构随着建筑的地面、平台慢慢升起，就像是抽象派画家笔下的火山喷发。那地表岩层似的“蘑菇”在阳光下，因本身的色彩，其轮廓显现出熔化的样子。由于采用的是钢丝网水泥结构，“蘑菇”可以黏着地面。室内的楼梯通向中心部分，而垂直部分连接着更衣室、平台及咖啡馆，看起来像个漂浮的飞碟。也正是这样的结构，使得观众可以观览全景，更好地欣赏足球比赛。

Barcelona City Hall

巴塞罗那市政厅

Project Location: Barcelona, Spain
Architectural Design: MiAS Arquitectes
Completion Year: 2010
Photography: Adria Goula
Building materials: Metal

项目地点：西班牙巴塞罗那
建筑设计：MiAS Arquitectes 建筑事务所
竣工时间：2010 年
摄　　影：Adria Goula
建筑材料：金属

Reproduce the Charm of the City 再现城市魅力

Project Description

The project meant a chance to go back to the neighborhood in an interested manner.It was no longer a trip down there for fun, to discover its people, its bars, its smell... but rather a survey of the place with the object of identifying what would enable us to reveal its qualities and to describe it accurately for the purposes of a project. In fact the market has always been an element of social cohesion in the neighborhood, a landmark, sometimes almost secret and visible only to its inhabitants.

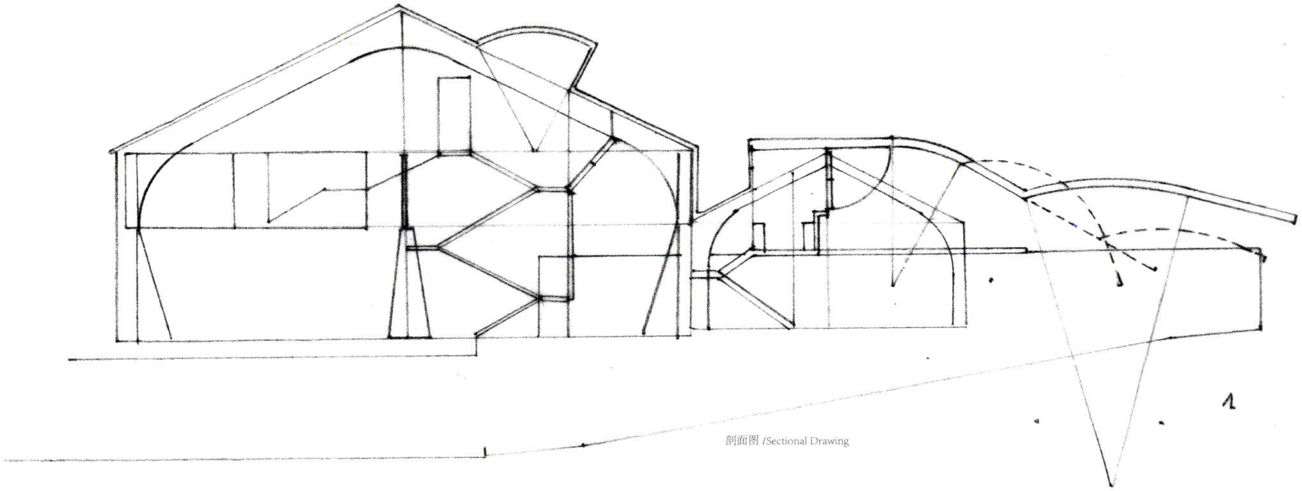

剖面图 /Sectional Drawing

Design Details

The market seeks to form a part of the neighbourhood, its urban fabric, and is redirected towards the squares front and rear-formerly no square existed, and the bays that made up the market crossed. The new metal figures create new market spaces, not touching the ground, but suspended from the old structure, not in a real manner, since the two structures, the old and new, never really overlap structurally, rather they do so in a false equilibrium.

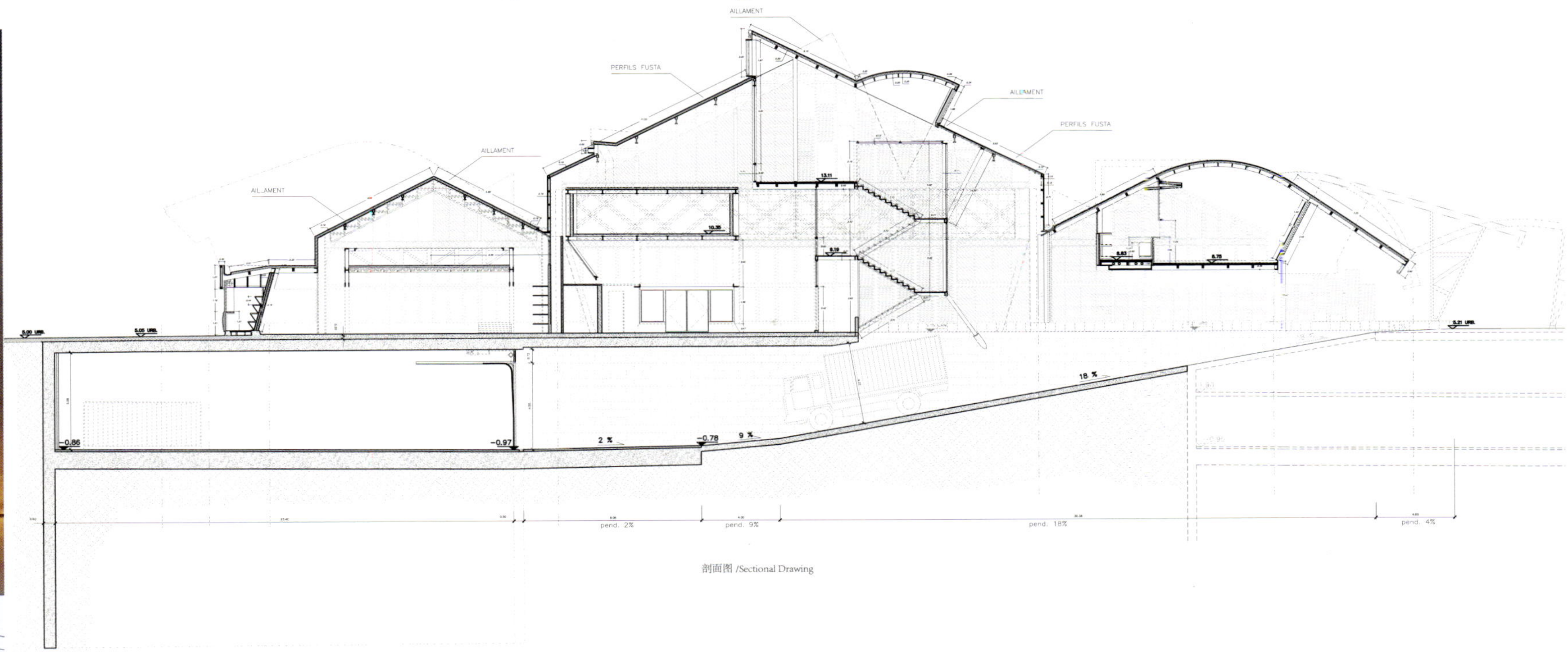

剖面图 /Sectional Drawing

项目介绍

该项目的落成意味着设计师用一种有趣的方式让人们重返邻里。人们不再因游玩而来到这里，人们怀念的不再是这里的人们、酒吧以及这里的味道，而是对这里的装饰进行研究，从而得出何种装饰能反映出城市的形象和魅力的结论，并精确地描述出它的外形，以协助新项目的落成。事实上，该市场已成为社会凝聚力的一个元素、一个里程碑，有时几乎是隐形的，而只有附近的居民才能看得见。

设计细节

该市场尝试着成为邻里的一部分，成为城市的结构，海湾正对着广场的前面和后面，使市场呈交叉型。新的金属轮廓创造了市场空间，不与地面接触，不以一种真实的方式，仅悬浮在旧建筑上。因为这两栋新旧建筑，从不在结构上重合，而是给人一种假重合的感觉。

CARNISSERIA
REGALOS Y HOGAR

Embotits
POLLERIA
REGALOS Y HOGAR

ROCKWOOL
ROCKWOOL

The Pod Pavilion

陈列室 + 庇护所

Project Location: Malaysia
Architectural Design: Studio Nicoletti Associati
Project Area: 2010
Completion Year: 800 sq.m.
Client: MKH Group
Photography: H. Lim Ho
Building materials: Steel, Aluminium Panels

项目地点：马来西亚
建筑设计：Studio Nicoletti Associati 建筑事务所
项目面积：2010 年
竣工时间：800 平方米
开 发 商：MKH 集团
摄　　影：H. Lim Ho
建筑材料：钢材、反光铝板

A Landmark in Jaya City 再也市的新兴地标

Project Description

In the area of Petaling Jaya, west of Kuala Lumpur, a great urban development is under way for the establishment of a new urban centre. As a landmark for this area, the developer wanted to host his on-site offices and sale's showroom in an iconic pavilion that would reflect the spirit and the architectural style of the whole development.

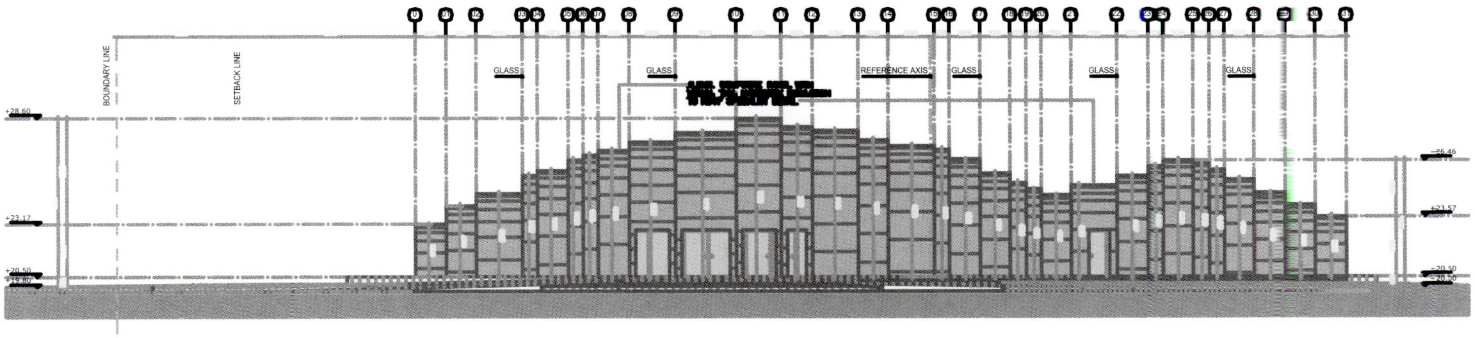

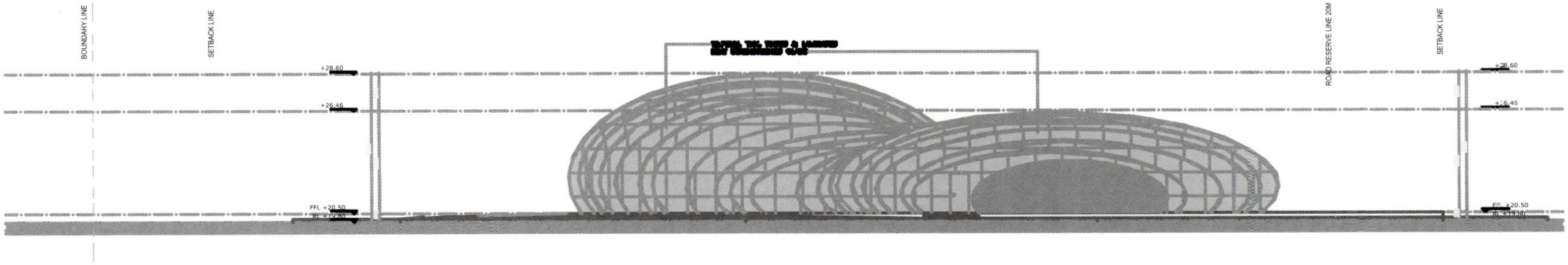

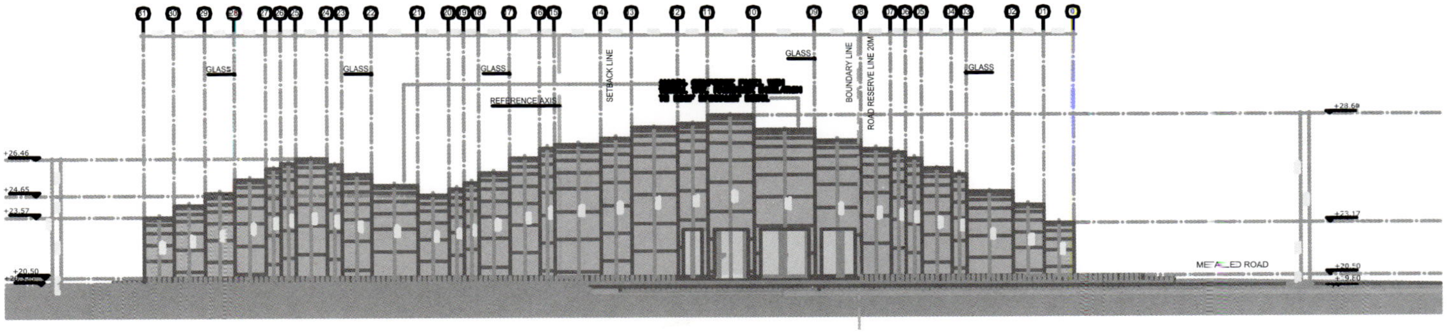

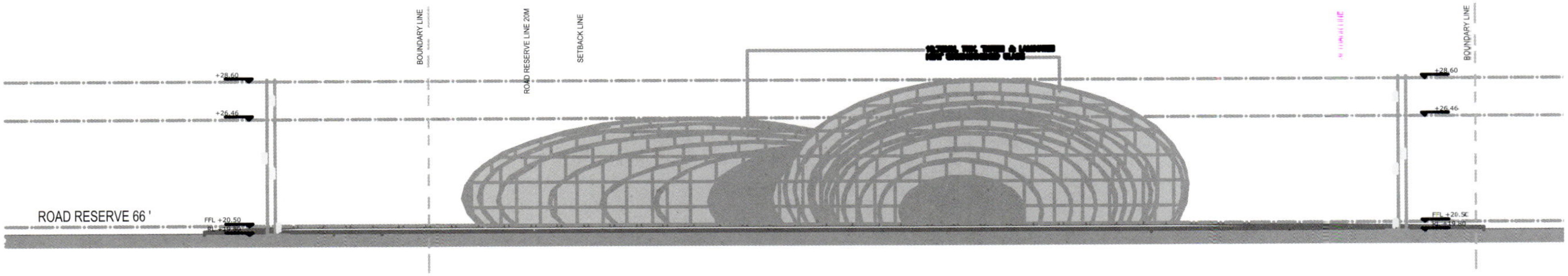

Design Details

Water droplets in nature were the inspiration for"The Pod"pavilion structure creating a dynamic spherical form resulting in a primitive building archetype with a modern twist. The round and soft shape of The Pod is formed as a series of elliptical sections of variable widths and heights. Slithers of windows bring natural daylight into the spaces below. Internally the pavilion is divided into two parts: one zone is dedicated to the office area and the other contains the main showroom. The building appears to be sliced diagonally into a series of ribbons which wrap up and over .It creating a dynamic space within, forming a layered protective shell. The structure is fabricated from tubular steel members with the exterior skin made of spectrally reflective aluminium panels. The exterior skin color shades changes depending on the reflection of the sun, dynamically. The Pod is surrounded by greenery with a reflective water pool wrapping around the edges that will contribute to the bioclimatic behavior and wellness of its inhabitants.

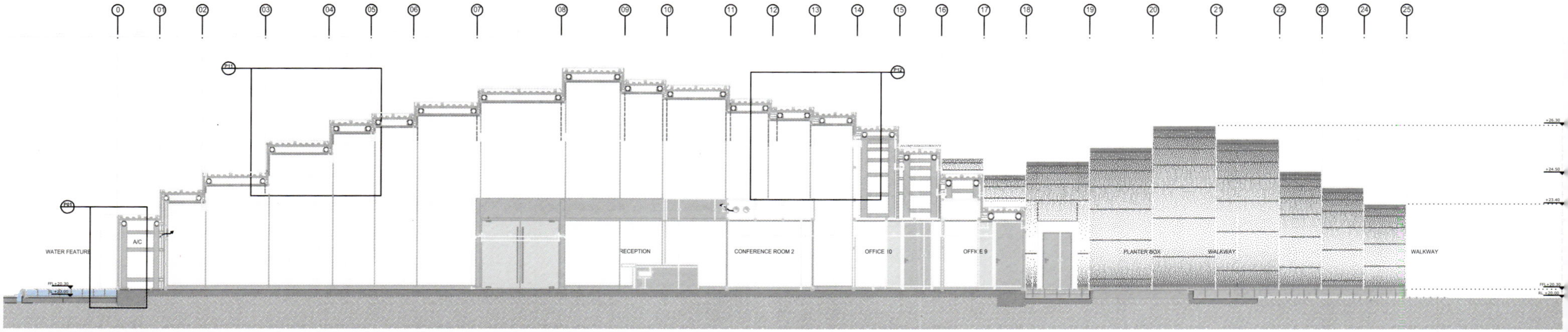
WATER FEATURE
A/C
RECEPTION
CONFERENCE ROOM 2
OFFICE 10
PLANTER BOX
WALKWAY
WALKWAY

项目介绍

该项目位于吉隆坡以西的八打灵再也市。为了建立一个新的市中心，一项伟大的城市发展正在进行之中。作为该地区的里程碑，开发商要求将自己的办事处和销售展厅设计在一个能反映该企业的精神和建筑形式的标志性建筑里。

设计细节

自然界中的小水滴便是该项目的设计灵感，一个充满活力的球形结构，再现了饱含现代风味的原始建筑形态。该项目圆润、柔软的外形形成一系列变换宽度和高度的椭圆形部分。滑动窗口为室内引入了充足的光线。该项目的室内分为两个部分：一部分是办公区，另一部分主要以展览为主。它沿对角线方向形成一条条色带，设计出动感空间，形成一个分层的保护壳。该项目采用钢材，室外则采用反光铝板，外表皮颜色的深浅由太阳光的反射而呈现出来。项目由一片绿色植物和一个反光水池围绕在它的周围，从而产生生物气候学现象，为居民营造一个健康的环境。

PJCC

PJCC

Monument to Pedro Almodovar

佩德罗·阿莫多瓦纪念碑

Project Location: Spain
Architectural Design: Enproyecto Arquitectura
Completion Year: 2009
Photography: Ricardo Santonja, Emilio Valverde
Building materials: Dry Self-Compacting Concrete

项目地点：西班牙
建筑设计：Enproyecto Arquitectura 建筑事务所
竣工时间：2009 年
摄　　影：Ricardo Santonja, Emilio Valverde
建筑材料：干拌自密实混凝土

A Spanish Public Building　西班牙公共建筑

Project Description

The monument is situated in the home town of Pedro Almodóvar, Calzada de Calatrava, in the heart of La Mancha. The competition asked about to link the birthplace of Almodóvar with his cinematographic work, strongly market with the presence of the culture and the landscape of his land. The main idea of the project was, in consequence, to honor La Mancha, enclosing in the monument the flat skyline of this land. The landscape remain captive into the monument, like it was a frame of an Almodovar´s movie.

Design Details

The stepped form of the structure is inspired in a camera focus, as well as its shape, resolving at the same time the accessibility of the monument to be used as a meeting point, a belvedere of the landscape, a place for concerts or the Almodovar's awards stage that the town concede every year. The park where it is situated forms a little hill, that let the monument to be seen from the road arriving to the town and to have a big prominence. The red colour is a wink to Almodóvar, who recognizes himself with this colour. The walls are textured wih shapes that remains his films.

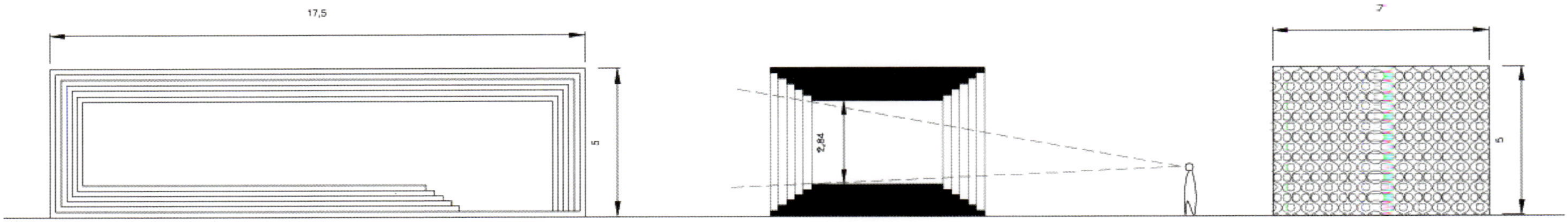

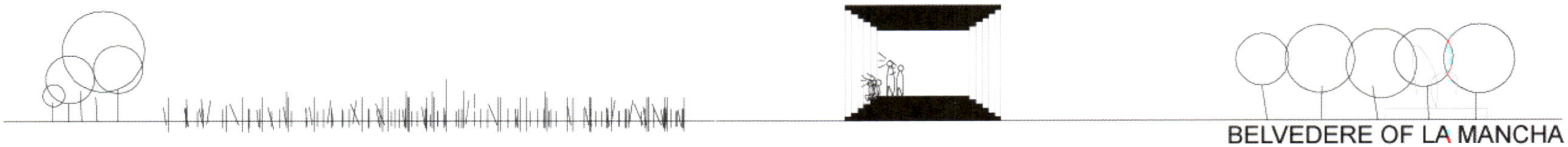

BELVEDERE OF LA MANCHA

项目介绍

该纪念碑位于佩德罗·阿莫多瓦的故乡拉曼查的卡尔萨达德卡拉特拉瓦小镇。该设计要求将阿莫多瓦的出生地与他的电影工作结合起来，着力表现这片土地上的文化和景观。该纪念碑的意念是向这片土地致意，符合平坦的天际线，让纪念碑成为阿莫多瓦的取景框。大地景观成为纪念碑捕捉的画面。

STAGE FOR ALMODOVAR´S AWARDS / CONCERTS / THEATRE

OPEN SUMMER CINEMA

设计细节

该项目梯形结构的灵感来自于相机的焦点，以及它的形状，同时纪念碑还用作集会场所、观景楼、音乐会或是阿莫多瓦的奖品台。公园形成了一个小山丘，让人们从路上便可见高挺的纪念碑。红色代表着阿莫多瓦，他用红色来辨别自己。墙壁设计成织纹型，以纪念阿莫多瓦出演过的电影。

Landscape Embellishment

景观装饰

The Tree
"大树"

Project Location: Singapore
Architectural Design: FARM
Completion Year: 2009
Photography: Jeremy San
Building materials: Steel, LED light tubes, Timber

项目地点：新加坡
建筑设计：FARM 建筑公司
竣工时间：2009 年
摄　　影：Jeremy San
建筑材料：钢材、LED 灯管、木材

A Magic Fort With History　历史古堡的神奇色彩

Project Description

When designers think about the National Museum, association with another grand old dame comes to mind. It is the huge old Banyan tree that sits majestically on the museum's front lawn, seemingly holding fort to a place full of histories, stories and magic. Across various cultures, the Banyan tree has always been a symbol of sanctuary where one finds peace of mind and harmony.

Design Details

The Tree will be made out of a series of varying sizes of rectangles and squares frames, rising from the ground and interlocking in a 3-dimensional space, forming a semblance of the Banyan Tree. Like an abstract modernist line painting, it will invite the viewers to walk around and explore this giant structure. At night, with the addition of LED light tubes, the Tree glows magically like a huge lantern. Its intricate web of structures traces lines in the night sky. As one stands under it and looks up, these lines of light would coalesce into imagined constellations of faraway stars.

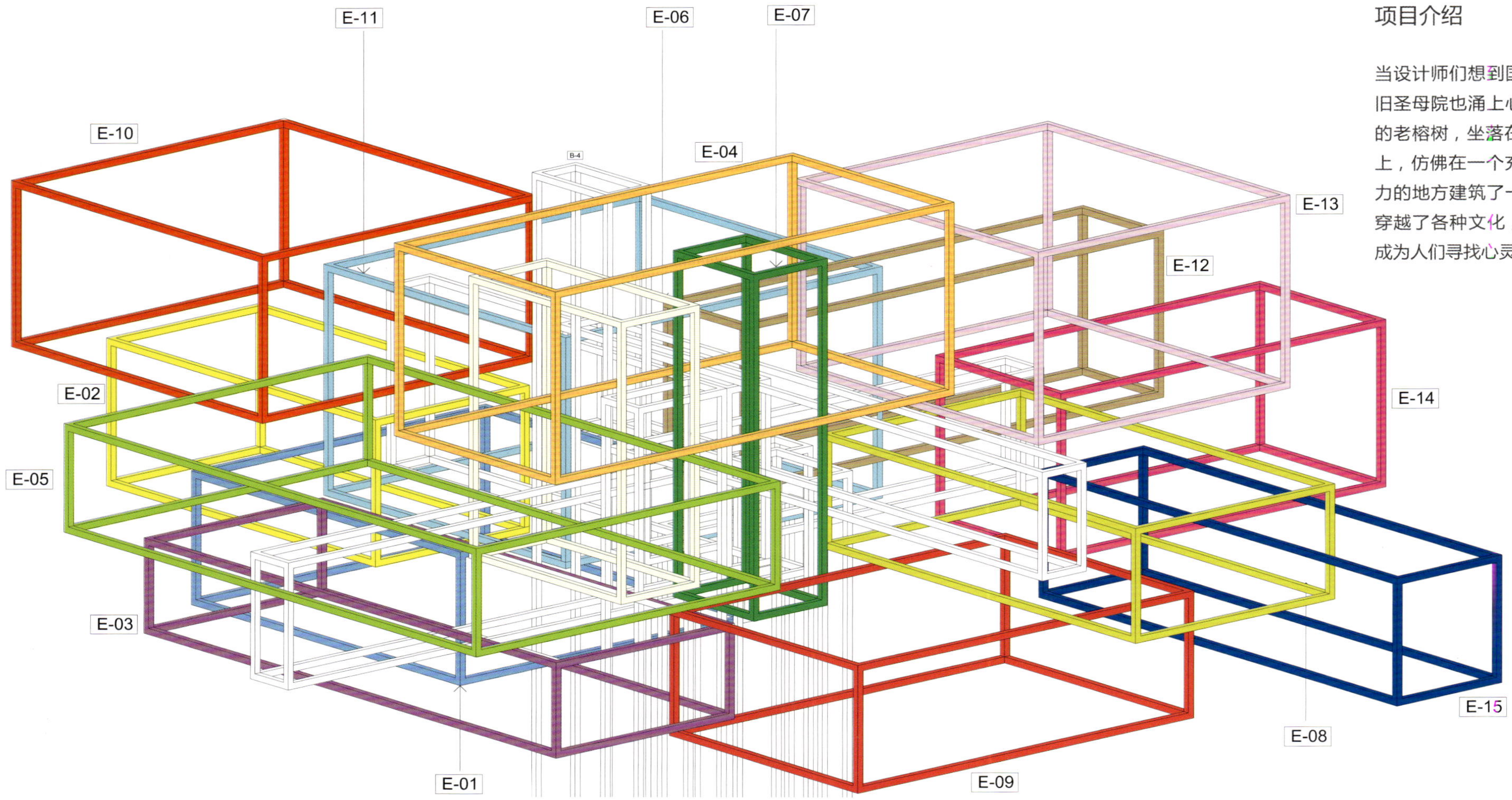

项目介绍

当设计师们想到国家博物馆，另一个旧圣母院也涌上心头。这是一棵巨大的老榕树，坐落在博物馆前面的草地上，仿佛在一个充满历史、故事和魔力的地方建筑了一个堡垒。这棵榕树穿越了各种文化，成为圣殿的象征，成为人们寻找心灵和谐的地方。

BOX NO.	DIMENSION (MM)	FR. GROUND	WELD NEXT TO
E-01	3000 X 2800 X 1050 (HT)	3200	D-04 : E-03
E-02	3000 X 1470 X 1150 (HT)	3800	D-04 : E-11
E-03	4600 X 1700 X 750 (HT)	3600	E-01 : E-06 : E-09
E-04	3800 X 1800 X 1500 (HT)	5700	D:02 : E-05
E-05	4600 X 2900 X 850 (HT)	4850	D-03 : E-04 : E-06 : E-10
E-06	1700 X 800 X 2200 (HT)	4350	B:03 : E-03 : E-05
E-07	900 X 700 X 2700 (HT)	4150	D-02 : E-09

BOX NO.	DIMENSION (MM)	FR. GROUND	WELD NEXT TO
E-08	3500 X 1900 X 900 (HT)	4400	D-02 : E-09 : E-13 : E-15
E-09	3200 X 2100 X 1000 (HT)	3400	E-03 : E-07 : E-08
E-10	2800 X 2600 X 1300 (HT)	5350	D:04 : E-05 : E-11
E-11	3020 X 2750 X 1800 (HT)	3850	B-04 : D-04 : E-02 : E-10
E-12	4000 X 900 X 1300 (HT)	4450	D:01 : E-13
E-13	2700 X 2400 X 1500 (HT)	5250	E-08 : E-12 : E:14
E-14	3100 X 1400 X 1300 (HT)	4000	D-01 : E-13 : E-15
E-15	4000 X 1000 X 1000 (HT)	3750	E-08 : E-14

设计细节

“大树”的建筑采用许多大小不一的矩形，从地面往上堆积，环环相扣，塑造出一个三维空间，形成一棵榕树。像一幅抽象的现代线性绘画，吸引着参观者走近观赏这个巨型建筑。夜幕降临时，在 LED 灯的映衬下，“大树”上闪耀着的灯光像一个大大的灯笼，它错综复杂的结构追寻着夜晚的高空。站在树下，抬头仰望便可见树上的线条与满天繁星合为一体。

Talud de la calle Jesús Galíndezy Plaza Pau Casals
保罗卡萨尔斯广场

Project Location: Bilbao, Vizcaya , Spain
Architectural Design: ACXT Architects
Completion Year: 2009
Client: Ayuntamiento de Bilbao
Photography: Aitor Ortíz
Building materials: Triangular Plane

项目地点：西班牙毕尔巴鄂
建筑设计：ACXT 建筑事务所
竣工时间：2009 年
开 发 商：Ayuntamiento de Bilbao
摄　　影：Aitor Ortíz
建筑材料：三角形平板

Bilbao's Physical Barrier 毕尔巴鄂的自然屏障

Project Description

Aim of the project is to consolidate the embankment. Recover this derelict area, which acts like a physical barrier and disintegrating element in the city. Designers propose to shape the embankment by using inclined planes of different materials, which reveal their strange topography to the city. The triangular planes are formed by different materials: the existing rock, vegetation of different colours, concrete in those areas which required consolidation and light, reconstructing their silhouette at night. To create connecting elements between the top and bottom levels and to lessen the impact of the embankment as a physical barrier in the city, they create a horizontal platform to take advantage of the height of this area in the city.

Design Details

All the elements in the project form part of this recreated topography: the stairways, the sitting areas, the public toilets. All these elements are included within a single fold. Following the embankment project, designers were commissioned to develop the Plaza de Pau Casals project and create a children's play area over a former electricity substation. All these locations are situated on steep slopes next to the embankment. It has been necessary to treat these spaces in accordance with the rules of geometry. Designers continue using triangular planes. In the children's playing area, the planes are soft-of grass, rubber, flowers. There are built-in toboggans. The old substation is enveloped in a fold of wooden slabs that convert it into part of the topography. In the Plaza de Pau Casals, the folds envelop a retaining wall and generate a space edged by gentle slopes where formerly there had been a crossroads; planting planes between the existing trees in the old central reserve; planes of coloured concrete and planes of water to sit next to.

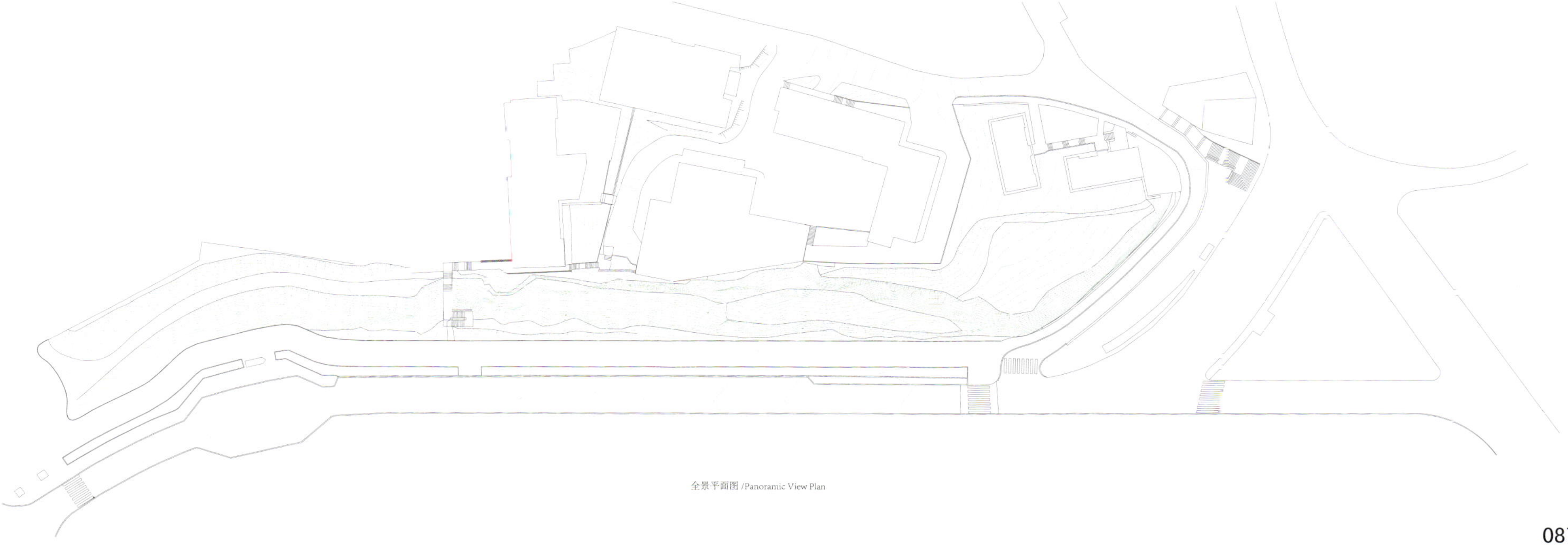

全景平面图 /Panoramic View Plan

项目介绍

该项目的设计是为了巩固河堤，使这片被遗弃的土地恢复生机，成为城市的一道自然屏障和风景线。设计师想采用不同材质的倾斜板作材料，为城市增添奇异的地形。三角平板的设计采用不同的材料，如岩石、色彩不同的植被，并在那些需要巩固的地方使用混凝土和光源，在晚上重现它们的轮廓。为了连接顶部和底部，减少河堤作为自然屏障的影响，设计师们充分利用这一地区的海拔高度，创造了一个横向平台。

aspes

设计细节

该项目所有的元素形成了这块重新构建的地形，例如楼梯、休息区和厕所。所有元素的设计只包括一道折痕。随着项目的进展，开发商要求设计师们在发电站上面设计一个名为保罗卡萨尔斯的广场和一个供孩子们游玩的场所。所有游玩的场所都位于河堤陡峭的山坡上，但它们的设计必须符合几何规则。设计师们不断使用三角形平板。孩子们游玩区的平板都是柔软的，用草坪、橡胶和花平铺在面板上。旧发电站用一些木砖围绕起来，使它成为地形的一部分。在保罗卡萨尔斯广场，这些折痕平板围绕着墙壁，并将以前的十字路口用缓坡形成一些空间。在中央地区，有一些种植区位于树木间，还有彩色混凝土平板和水池。

Giant Digital Origami Tigers

巨型数字折纸虎

Project Location: Australia
Architectural Design: LAVA
Completion Year: 2010
Building materials: Aluminium, Barrisol, LED lighting

项目地点：澳大利亚
建筑设计：LAVA 设计公司
竣工时间：2010 年
建筑材料：铝、巴力软膜、LED 灯

A Celebration to Chinese New Year　恭贺中国虎年

Project Description

Two giant digital origami tigers playfully celebrate Chinese New Year on the forecourt of Customs House, Sydney from 11 February—14 March 2010. The crouching digital tigers combine ancient lantern making methods with cutting edge digital design and fabrication technology, bringing east and west together through tradition and innovation. The big cats are collaboration between multinational architectural practice Laboratory for Visionary Architecture (LAVA) and Customs House to mark the lunar year of the tiger and raise awareness about the endangered status of tigers.

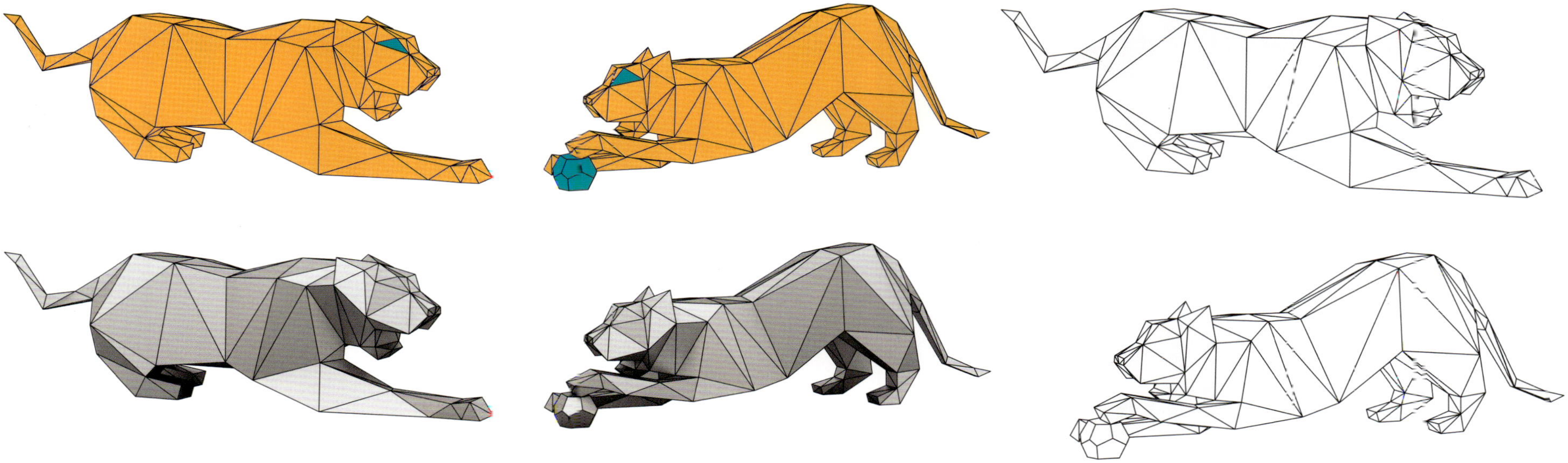

Design Details

The tigers are inspired by "zhezhi", a Chinese term for paper folding, more popularly known by its Japanese name "origami". Traditional Chinese lantern makers in Sichuan province, where lantern making has been taking place for over 300 years and local artisans were consulted. The Chinese or Lunar New Year is the most important Chinese holiday.The Chinese lunar astrological sign is the metal tiger in 2010. The tigers are the size of a boat at 2.5 metres high and 7 metres long yet weigh only 200 kg. It uses fully recyclable materials, aluminium and barrisol, a new light weight reusable stretch material. Pulsating low energy LED lighting brings the sculptures to life.

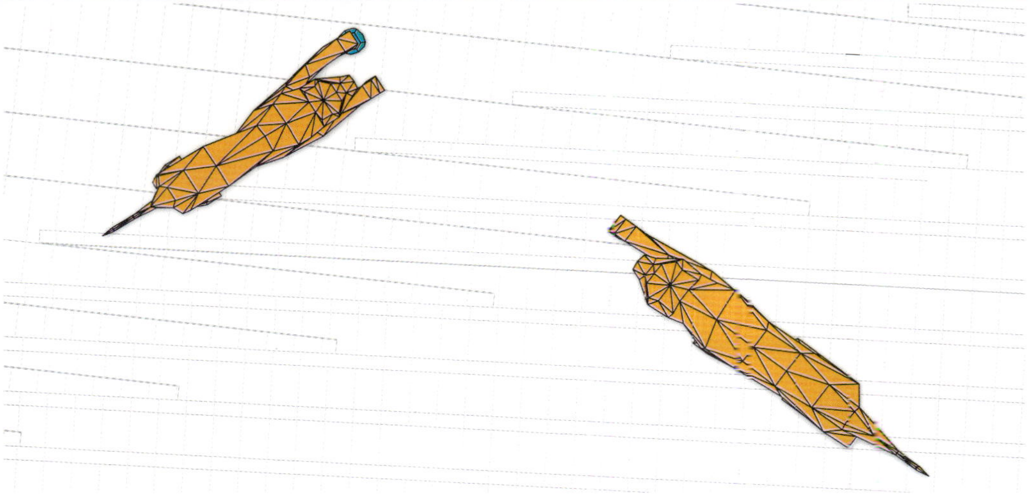

项目介绍

从 2010 年 2 月 11 日 ~3 月 14 日，有两个巨大的数字折纸老虎在悉尼海关大楼的前院里，调皮地庆祝着中国新年。蜷缩着的数字老虎的设计用尖端数字化方法和制造技术，并融合古灯笼的制作方法，通过传统与创新将东西方结合在一起。该项目的完成是由跨国梦幻建筑实验室和海关联合打造，以此来庆祝中国农历虎年，并提高人们对老虎濒临危机的认识。

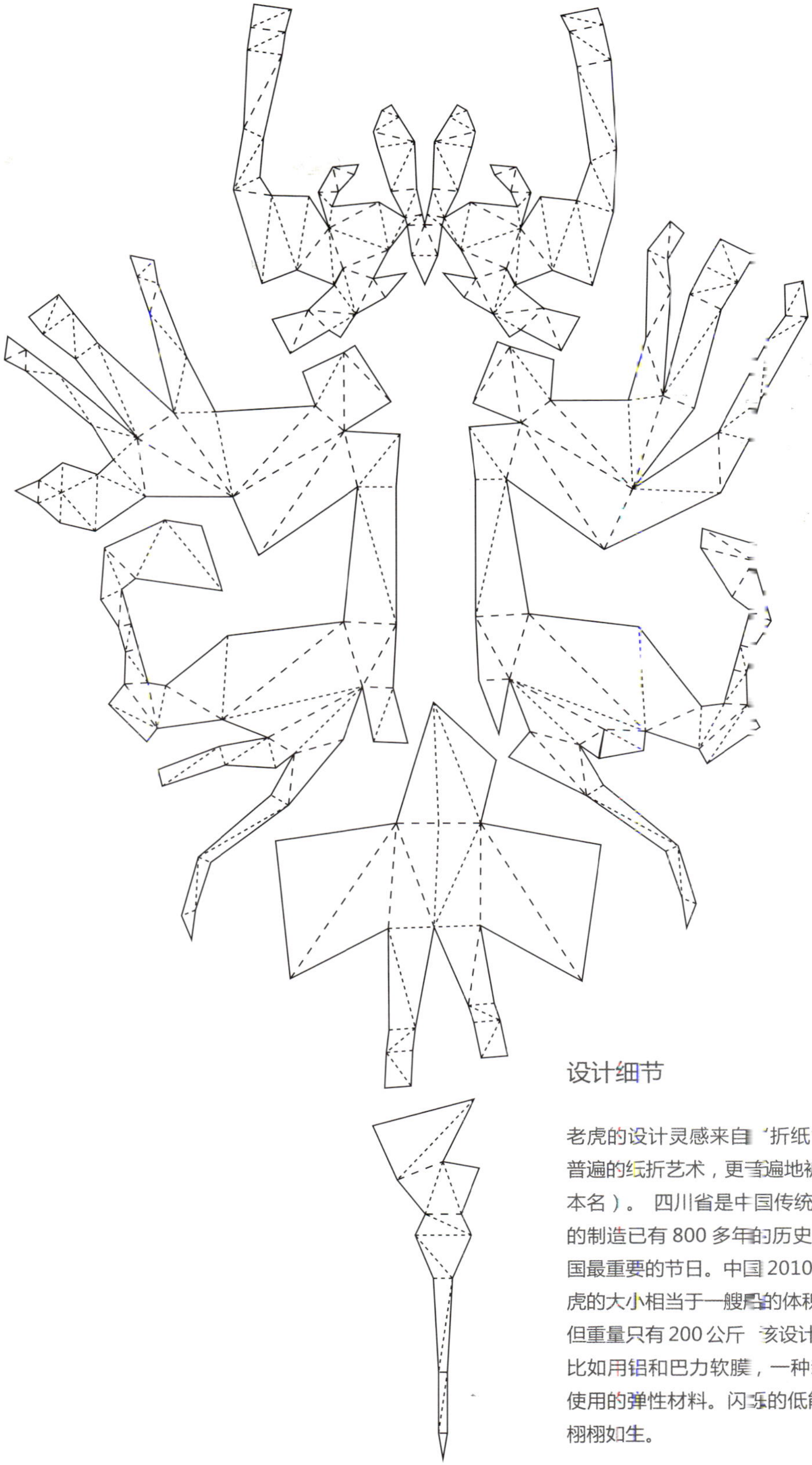

设计细节

老虎的设计灵感来自“折纸”，一种中国长期比较普遍的纸折艺术，更普遍地被称为“crigami”（日本名）。 四川省是中国传统灯笼的制造商，其灯笼的制造已有 800 多年的历史。春节或农历新年是中国最重要的节日。中国 2010 年农历生肖属相是虎，虎的大小相当于一艘船的体积，长 7 米，高 2.5 米，但重量只有 200 公斤。该设计充分利用可回收材料，比如用铝和巴力软膜，一种新的重量轻，且可重复使用的弹性材料。闪烁的低能 LED 灯将雕塑刻画得栩栩如生。

Vinaros Promenade
比纳罗斯长廊

Project Location: Vinaros, Castellon. Spain
Architectural Design: Guallart Architects
Completion Year: 2007
Client: Vinaros City Council
Photography: Laura Cantarella

项目地点：西班牙卡斯特利翁比纳罗斯
建筑设计：Guallart Architects 建筑事务所
竣工时间：2007 年
开 发 商：比纳罗斯市政府
摄　　影：Laura Cantarella

An Seashore Gallery 海滨广场

Project Description

Vinaros is the most northerly town of the Valencia region. During the 20th century, urban growth made it necessary to delimit the town's relation with the sea by means of a seafront promenade that subsequently became a pole of attraction for the tourist boom that Spain began to experience in the 1960s. In fact ,Vinaros did not undergo the processes of peripheral growth or the programmatic consequences of delocalization of its commercial activities typical in 1980s, or the massive influx of tourist hotels and franchises that occurred in neighboring towns. As such it presents a very interesting urban type, in many respects having jumped directly from the 1970s to the 21st century. Reform of the seafront promenade, as the interface between the centre of the town and the sea, offered a great opportunity for a public initiative to define the desired standards of urban quality for future growth. In fact the promenade is today a place of great urban vitality, occupied mainly by cars, which coexist with seafood restaurants, an open-air auditorium for orchestral and choral concerts and open-air film screenings, and the regular street markets and events throughout the year, frequented by people on their way to and from the beach.

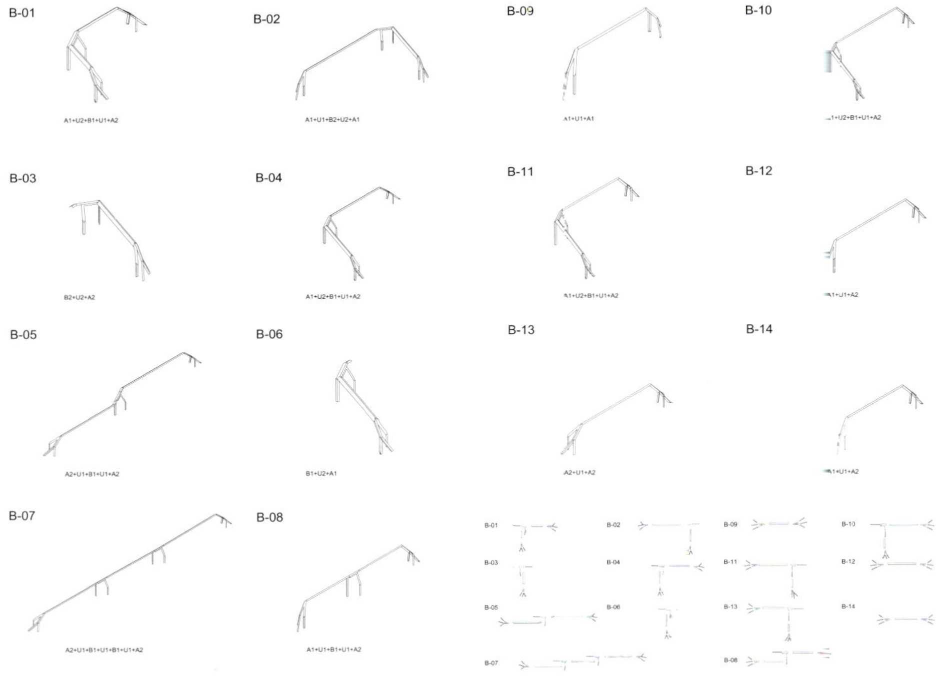
B-01
A1+U2+B1+U1+A2
B-02
A1+U1+B2+U2+A1
B-09
A1+U1+A1
B-10
1+U2+B1+U1+A2
B-03
B2+U2+A2
B-04
A1+U2+B1+U1+A2
B-11
A1+U2+B1+U1+A2
B-12
A1+U1+A2
B-05
A2+U1+B1+U1+A2
B-06
B1+U2+A1
B-13
A2+U1+A2
B-14
A1+U1+A2
B-07
A2+U1+B1+U1+B1+U1+A2
B-08
A1+U1+B1+U1+A2
B-01
B-02
B-09
B-10
B-03
B-04
B-11
B-12
B-05
B-06
B-13
B-14
B-07
B-08

Design Details

This car park will be connected with others in or under adjacent squares to create a real underground mobility network that allows cars to drive in from the outskirts to the centre of the town and park close to the beach and the seafront promenade. It was also decided to eliminate the concrete wall separating the beach from the promenade to enable the whole area to be perceived as a continuous space composed of a variety of materials, from the sand of the artificial beach to the cenia stone from a local quarry. Another significant decision was that the promenade, which at present has an irregular topography, should have a constant level that would set off its eight-hundred meter horizontal line against the natural line of the sea's horizon. This serves to resolve the difference in level between the beach and the promenade by means of a system of tiers that can be occupied in a variety of ways.

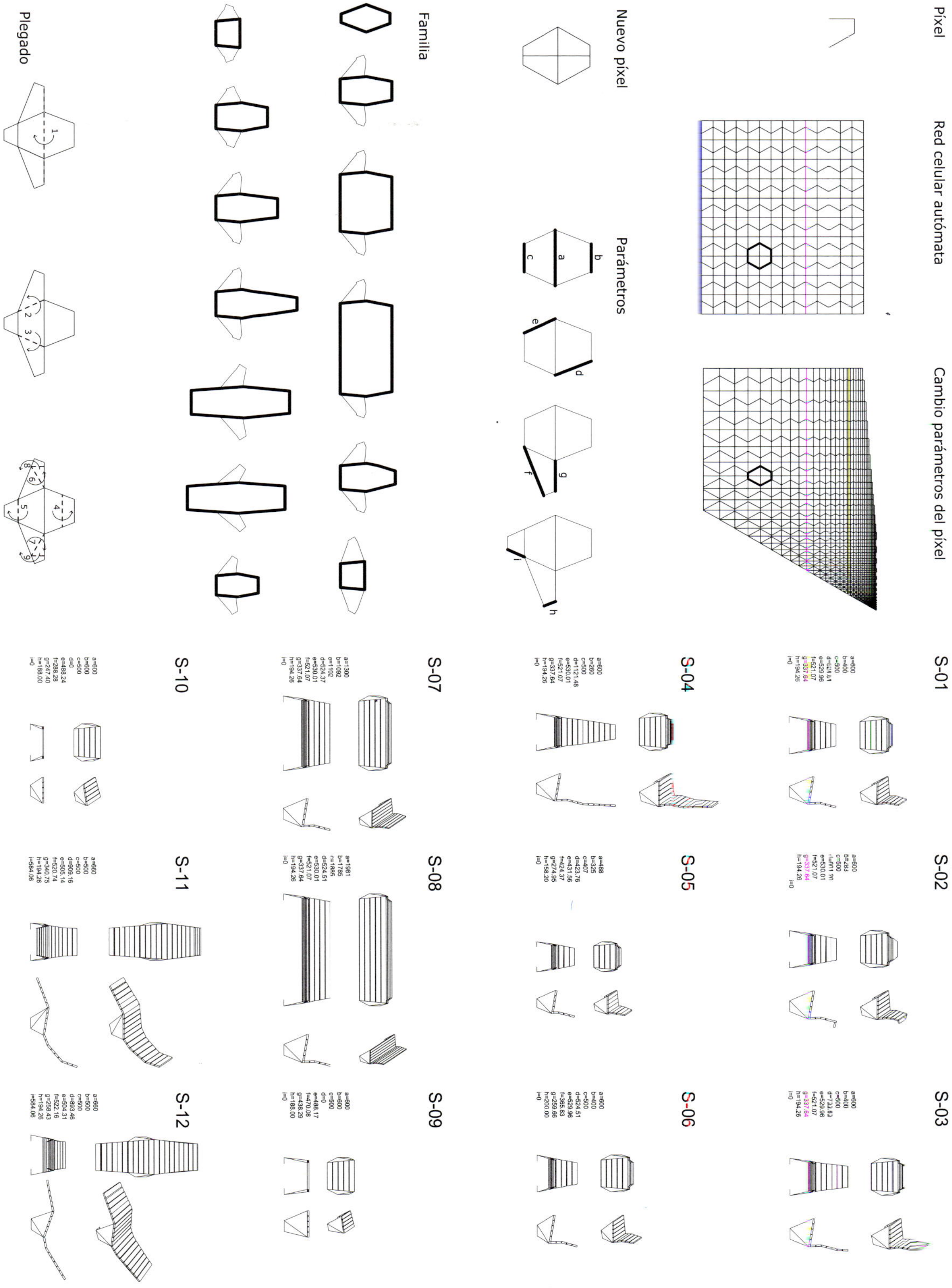

项目介绍

比纳罗斯是巴伦西亚地区最北端的城镇。在20 世纪期间，城市化增长使其必定要用海滨长廊将城镇与海洋的关系划分出来，它在20 世纪 60 年代的旅游热潮中曾发挥重要作用。事实上，比纳罗斯还未曾经历过其周长增长、计划性系列及 20 世纪 80 年代商业活动类型的不定领域，或大量旅游饭店的涌入和邻镇的特许经销酒店等。这样使它呈现出一个十分有趣的城市类型，从许多方面来说，它已从 20 世纪 70 年代直接跨跃到 21 世纪。海滨长廊的改革中，它作为镇中心和海洋的分界线，为公众提供了一个很好的机会来定义这座城市未来发展所需的质量标准。事实上，如今的海滨长廊只是一个融合伟大城市活力的地方，主要是由小汽车停车位、海鲜餐厅、交响乐团和合唱音乐会的观众席、露天电影银幕以及一年中的定期集市和活动所占据，海滨长廊时常有路人和海滩上的游客光顾着。

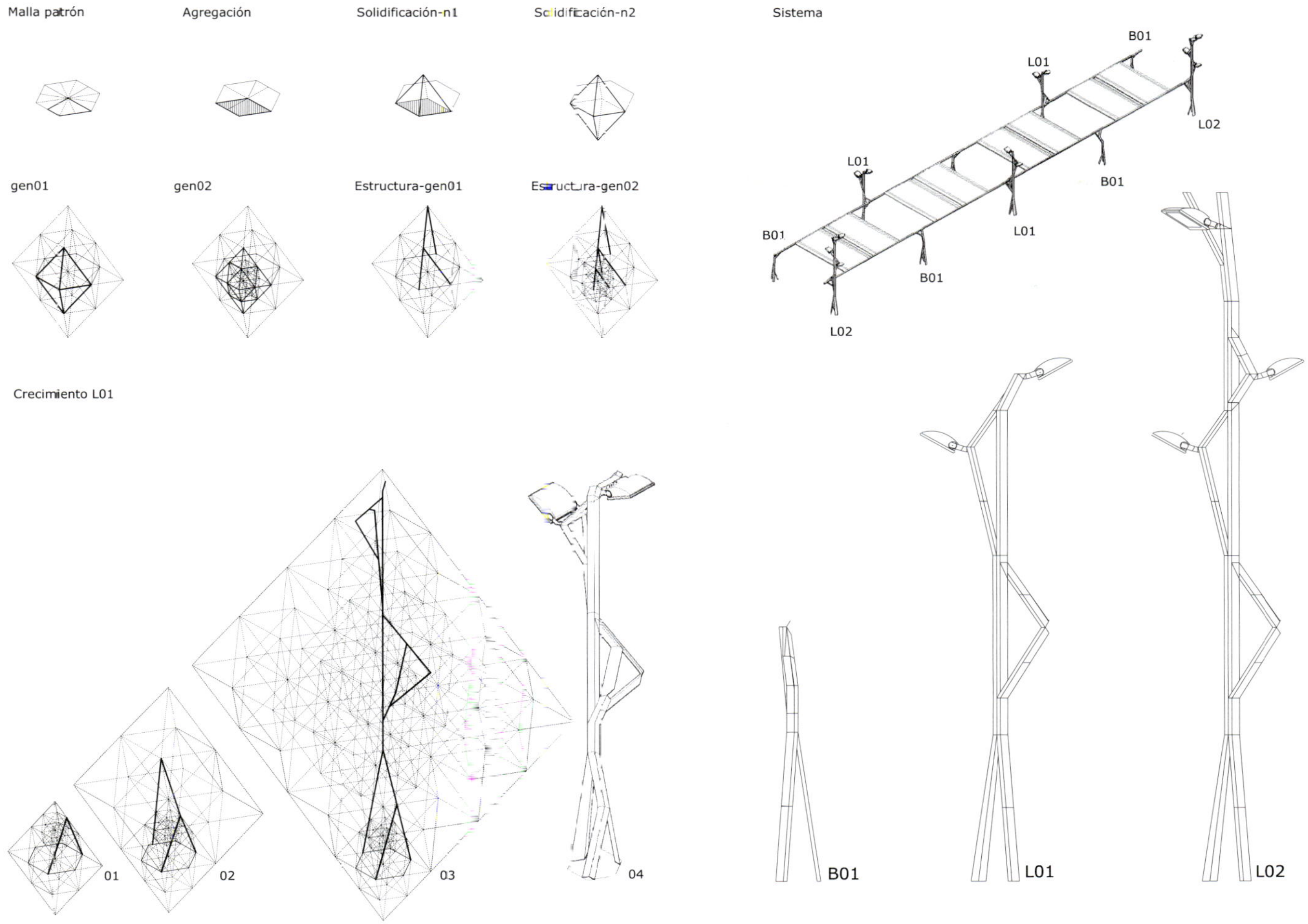

设计细节

设计师将停车场与相邻广场相连，创造一个真正的地下移动网，允许车辆从近郊驶入市中心，并停靠在靠近海滩和海滨长廊的位置。设计师决定撤除介于长廊和海滩之间的混凝土墙，这可以更方便地让人看到长廊上从人造沙滩到塞尼亚石头等所有建筑。另一个具有重要意义重要的决定是长廊的设计，由于它的不规则地形，因此也应该有一个不变的标准来衬托它 800 多米长的水平线和海上地平线。设计师用层层排列的设计方法，解决了长廊与海滩之间的不同层次的问题

2 Piers

2 架桥墩

Project Location: Sweden
Architectural Design: GORA art & Landscape
Completion Year: 2008
Building materials: Steel

项目地点：瑞典
建筑设计：GORA art & landscape 设计公司
竣工时间：2008 年
建筑材料：钢材

A Way to Soul 通往心灵深处的桥梁

Project Description

2 Piers is a solution that points out, enhances and intensifies the view. It also—by being something unexpected-creates focus in something vast. At the same time it is a fairly mild intrusion, clear-cut but mild. Two piers on a slope far above the water surface also create some mystery.

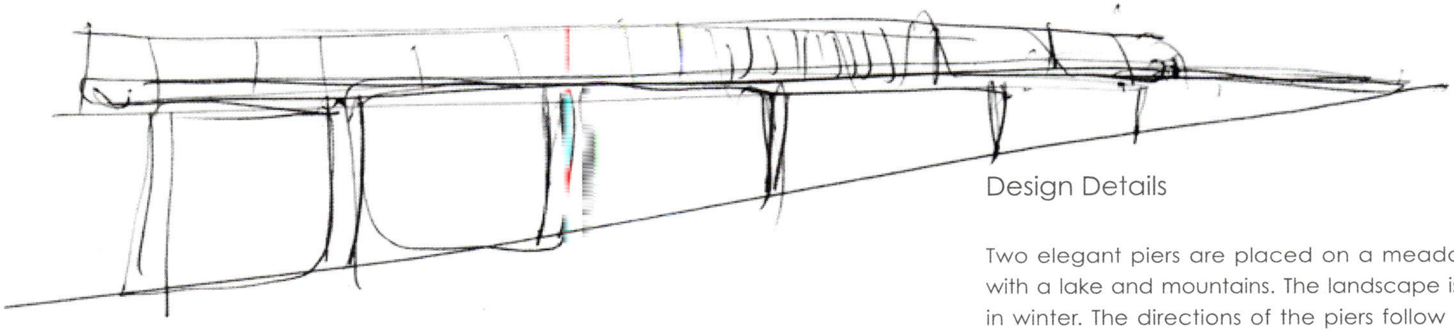

Design Details

Two elegant piers are placed on a meadow and jut out into a grand northern landscape with a lake and mountains. The landscape is green and blue in summer, grey, white and blue in winter. The directions of the piers follow lines that-if continued into thin air-will eventually intersect. One of them twenty and the other thirty meters long, they rest on steel pillars, and are made of larch and steel. They run horizontally, but the slope beneath creates a sense of ascent and adventure as one walks out on them. From certain angles the pillars seem to be walking towards the lake. The 2 Piers measure the slope like spirit levels, and provide a playful place for landscape contemplation.

项目描述

这两架桥墩的设计是为了提高和加强俯瞰风景的角度。它采用一些意想不到的东西，在广阔的空间里制造了许多焦点。这两架桥墩建在山坡上，腾驾于水面，带来一种奇妙的感觉。

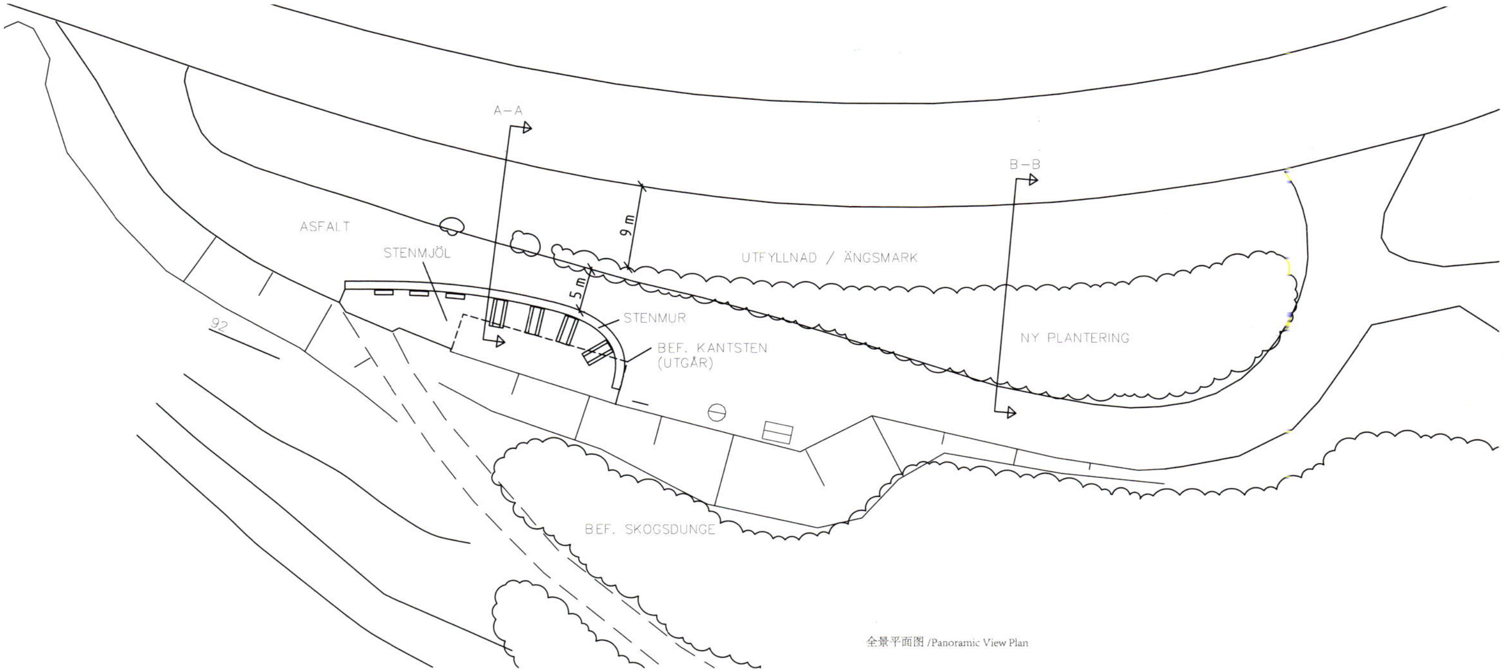

全景平面图 /Panoramic View Plan

设计细节

两架设计雅致的桥墩从草地向南面的湖和山脉延伸出去。夏日的景色是绿色和蓝色的，到了冬天，便是灰色、白色和蓝色了。桥墩的设计方向，如果继续延伸，最终将在空中相交。其中一架桥墩20米长，另一架30米长，采用钢柱做桥墩，他们横向穿梭，一旦走向桥的顶端，下面的坡度会给人一种腾空和冒险的感觉。从一定的角度来看，桥墩延伸的角度似乎是朝着湖面设计的。这两架桥墩的高度带给人心灵的宁静，还提供一个可供游玩赏景的地方。

Keelung Maritime Plaza

基隆海洋广场

Project Location: Barcelona, Spain
Architectural Design: Guallart Architect
Completion Year: 2009

项目地点：西班牙巴塞罗那
建筑设计：仲观设计顾问有限公司
竣工时间：2009 年

A Bustling Port 繁华的海港

Project Description

Keelung is the port of Taipei, the capital of the island of Taiwan. Located 30 km to the north of the capital, it is one of the most important container ports in Asia. Keelung has all the vitality of a major port, with one of the most bustling night-time markets in the Far East and an extensive and multifarious central commercial area adjoining the port. The city nevertheless bears the traces of rapid economic growth. Its principal transportation infrastructures—roads, railway lines, and the port itself—continue to limit the creation of public spaces quality in the downtown area.

Design Details

A city is defined internally and externally by the strength and quality of its central space. Designers' proposal offers the opportunity of creating a more powerful and complex urban centre than will come to constitute Keelung's symbolic space and serve to project an image of the city to the exterior. In the west zone of the area of intervention the three main transport infrastructures of the city converge. The solution that is envisaged for this space is not confined to embracing their existence as such, but seeks to facilitate the uses that these infrastructures induce and require, spatially organizing their satisfactory distribution and functional integration in the ways that are most convenient for the commuters and proposing cultural and leisure uses that will extend beyond the usual commuter timetable to avoid the risk of it losing its vitality and becoming marginalized.

In the west zone there are a variety of cultural, administrative and commercial facilities whose integration is problematic, because of the difficulties in the way of pedestrian movement. At the same time there is a need to generate new uses and functions that will reinforce the relational contents of the public spaces, in order to provide a qualitatively richer urban framework for the citizens. This entails the restructuring of the area of amenities to the east of the port and the incorporation of new functions. In spite of living from the sea and with the sea, the people of Keelung are confronted by major obstacles when it comes to enjoying it, in view of the impact of vehicular traffic on the seafront and the very limited space that has been reserved for pedestrian circulation. To overcome this obstacle designers propose a solution that not only improves the flow of pedestrian circulation but also generates a space that permits the creation of a rest space where people can enjoy contemplating the sea and the bustling activity of the port.

KEELUNG

MNAC

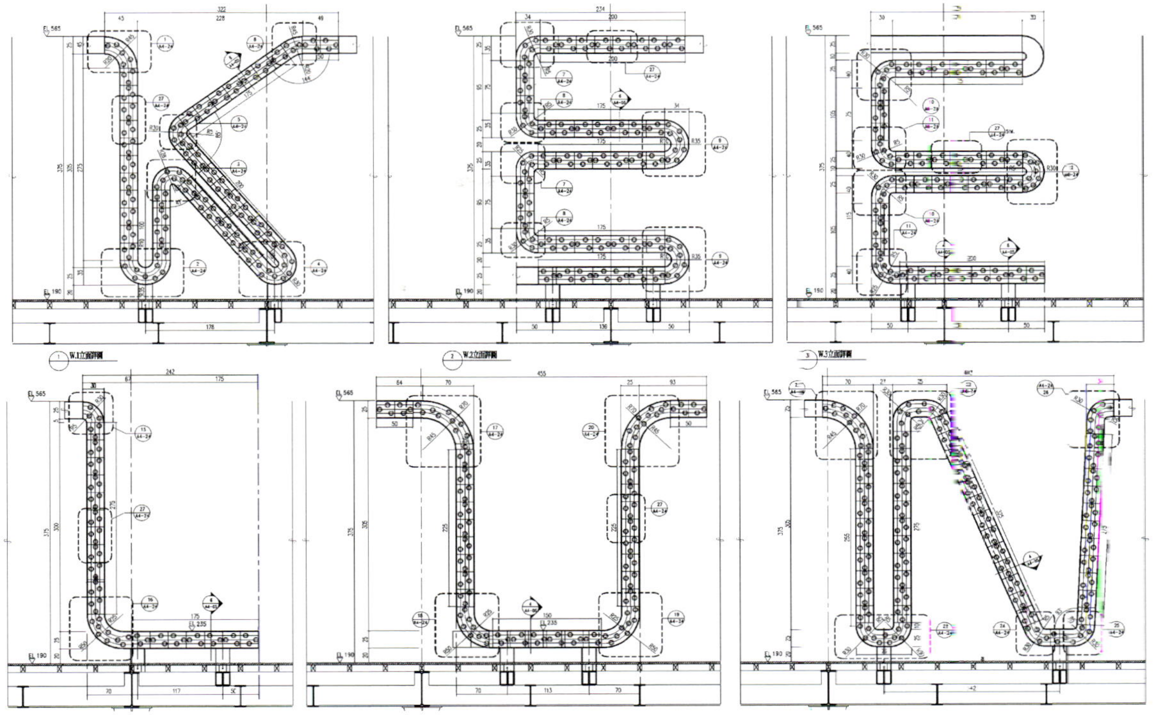

项目介绍

基隆是台北的一个港口。基隆海洋广场位于台北北面 30,000 米处，它是亚洲最重要的货柜港之一。基隆有着主要港口的所有活力，毗邻远东一个最繁华的夜市和广泛而多元化的中央商务区。这座城市仍然承载了它迅猛的经济增长足迹。其主要的运输基础设施包括陆路、铁路和水路，均限制了市区公共环境质量的创造。

设计细节

一座城市的繁华与否是用其中央空间内部和外部的力量和质量来权衡的。设计师提供了一个创造更多更强大和更复杂的市中心的机会，而不是构建基隆的象征性空间和塑造一个良好的对外城市形象。在该地区西区的干预领域，三个主要交通基础设施衔接在一起。设想的建筑方案是不仅限于包含他们自己的存在空间，而旨在达成这些基础设施所期望和要求的使用方法。在空间的设计上，以一种对乘客最便捷的方式，来实现他们满意的空间分布和集中性功能，并增添康乐及文化用途，超出平时的通勤时间表，以避免失去活力和被边缘化的危险。

在西区有各种文化、管理和商业设施，但将他们综合于一体是有困难的，因为行人的运动过程会使它所产生困难。同时，有必要让它产生新的用途和功能，这将加强公共场所关系的相关内容，以便为市民提供质量更丰富的城市框架。这就需要对东部港口这一舒适地区进行重建，并增加一些新的功能。除了出生在海边，生活在海边之外，当基隆人正准备享受这一美景的时候，却面临着一个主要障碍，那就是海边马路上拥挤的交通，只为行人留下非常狭窄的空隙。为了克服这一困难，设计师提出了一个解决方案，不仅提高了马路的顺畅度，同时也创造了一个新的空间作为公共休息场所，让行人在休息的同时还可以一览海景，以及熙熙攘攘的港口。

Metamorphosis
变形建筑

Project Location: Sweden
Architectural Design: GORA art & Landscape
Completion Year: 2008

项目地点：瑞典
建筑设计：GORA art & Landscape 设计公司
竣工时间：2008 年

Transformation Achieves Arts 变化成就艺术

Project Description

Designer chose the shiny material precisely because he wanted to capture a sense of the mutable, flowing state between other more solid states—it represents transformation. Metamorphosis is a sculpture that is linked to the major investment; the city of Linkping has made in biogas for the city's collective traffic. The gas is produced from agricultural waste which is converted through fermentation.

Design Details

The sculpture is an interpretation not only of the biological processes but also of mental change, new lines of thought and new associations. The surface reflects the surroundings, the plants, the sky, the clouds, the sun and the visitors. It both exists and does not exist at one and the same time; it hides in the reflections. It eludes you as fleetingly as the clouds' transitory accumulations of water vapour.

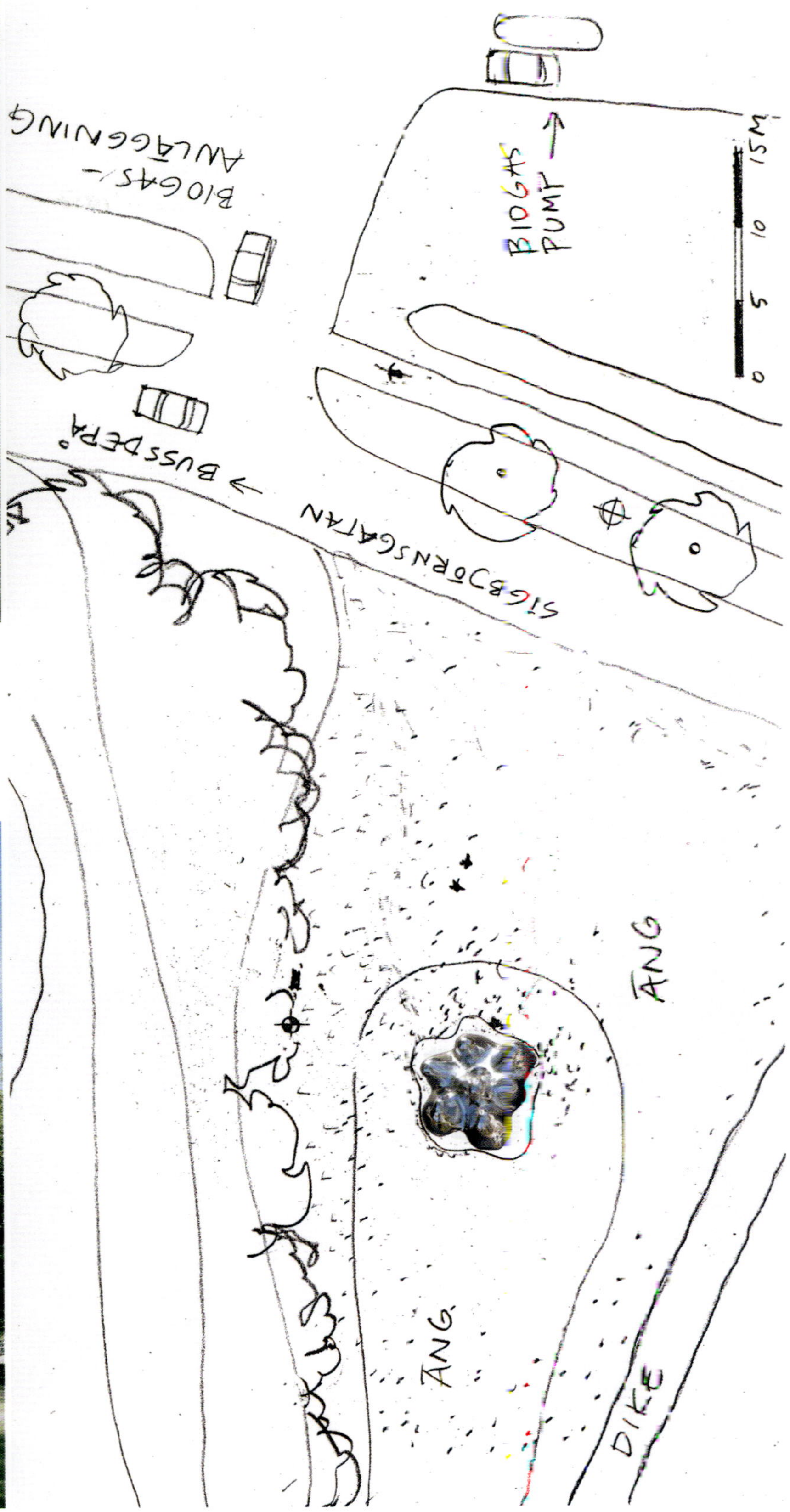

项目介绍

设计师特意选择了闪光材料来表现该项目多变的流动感。这是一个以沼气为原料制成的雕塑。气体是由农业废料发酵而成。

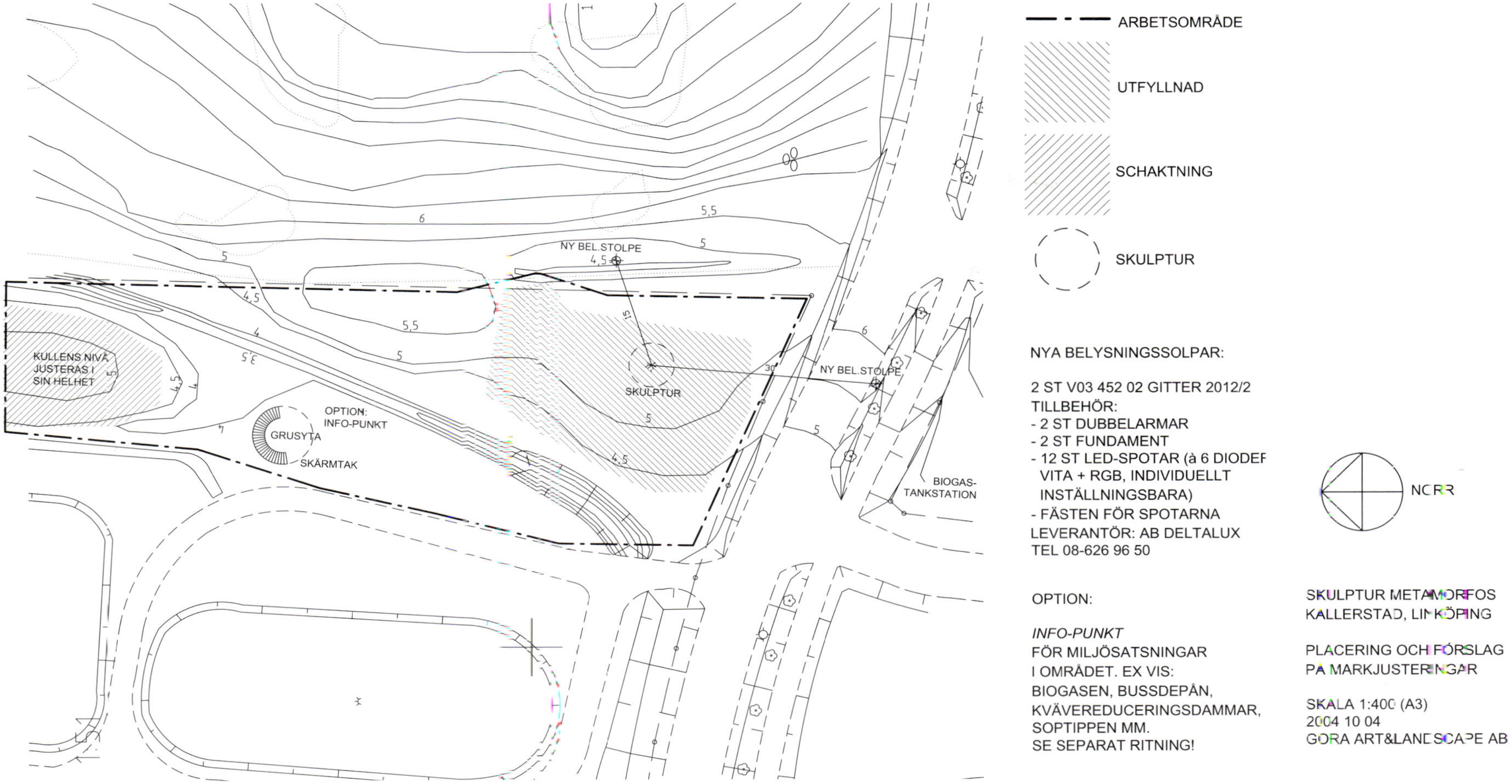

设计细节

该雕塑由生物加工而成，其设计理念是由新思想碰撞而成的。其表面映射着周围的景致——蓝天白云、阳光绿树，还有嬉戏的人们。这时雕像便变得时有时无，隐藏在倒影中。它像转瞬即逝的云流，又像是正要挥发的水蒸汽。

Forte Piano

钢琴

Project Location: Warsaw, Poland
Architectural Design: Quixotic Architecture Ltd.
Completion Year: During Construction

项目地点：波兰华沙
建筑设计：克维克索蒂克建筑事务所
竣工时间：在建

A Commemoration to the Great Master Fryderyk Chopin

纪念伟大的音乐家肖邦

Project Description

2010 marks the 200th anniversary of the birth of Fryderyk Chopin. This project was designed to commemorate one of the great Masters of romantic music; to celebrate the universality of music and to unite music lovers and enthusiasts. The location was carefully researched and selected as it is both in the very heart of the city of Warsaw, and lies parallel to the street where the Chopin Family lived. It is where young Fryderyk studied at the University of Warsaw, where he played organs in the Wizytek Church and where his Polish heart rests in the Church of Sw. Krzyz.

Design Details

The proposed monument is a giant piano keyboard. It is not only aesthetically pleasing but through its practicality is utilitarian and inclusive of the community irrespective of gender, age, nationality or creed. The installation pays subtle homage to Chopin through one of the greatest works he composed late in his short life. The keys F1, F2, B4, D5*, G5* are depressed. These are the first accords of his Barcarolle Fis-dur op.60. This detail gives the keyboard its own dynamic rhythm and will intrigue the passer-by. As a seating installation it welcomes the passer-by, offering a place for rest, reflection and contemplation. It can be a meeting place and also more formally can be used for open air piano concerts. The keyboard therefore will become part of a performance, something uniquely close to the audience.

项目介绍

2010 年是肖邦诞辰 200 周年。本项目的设计正是为了纪念这位伟大的浪漫主义音乐家和庆祝音乐的普遍性，并将音乐爱好者和狂热者联系起来。该项目的地理位置也是精心挑选的，位于华沙市中心，与肖邦家的街道平行。该项目所在的地方就是肖邦上过的华沙大学，他曾经在 Wizytek 教堂演奏过乐器，并在圣十字教堂里祷告。

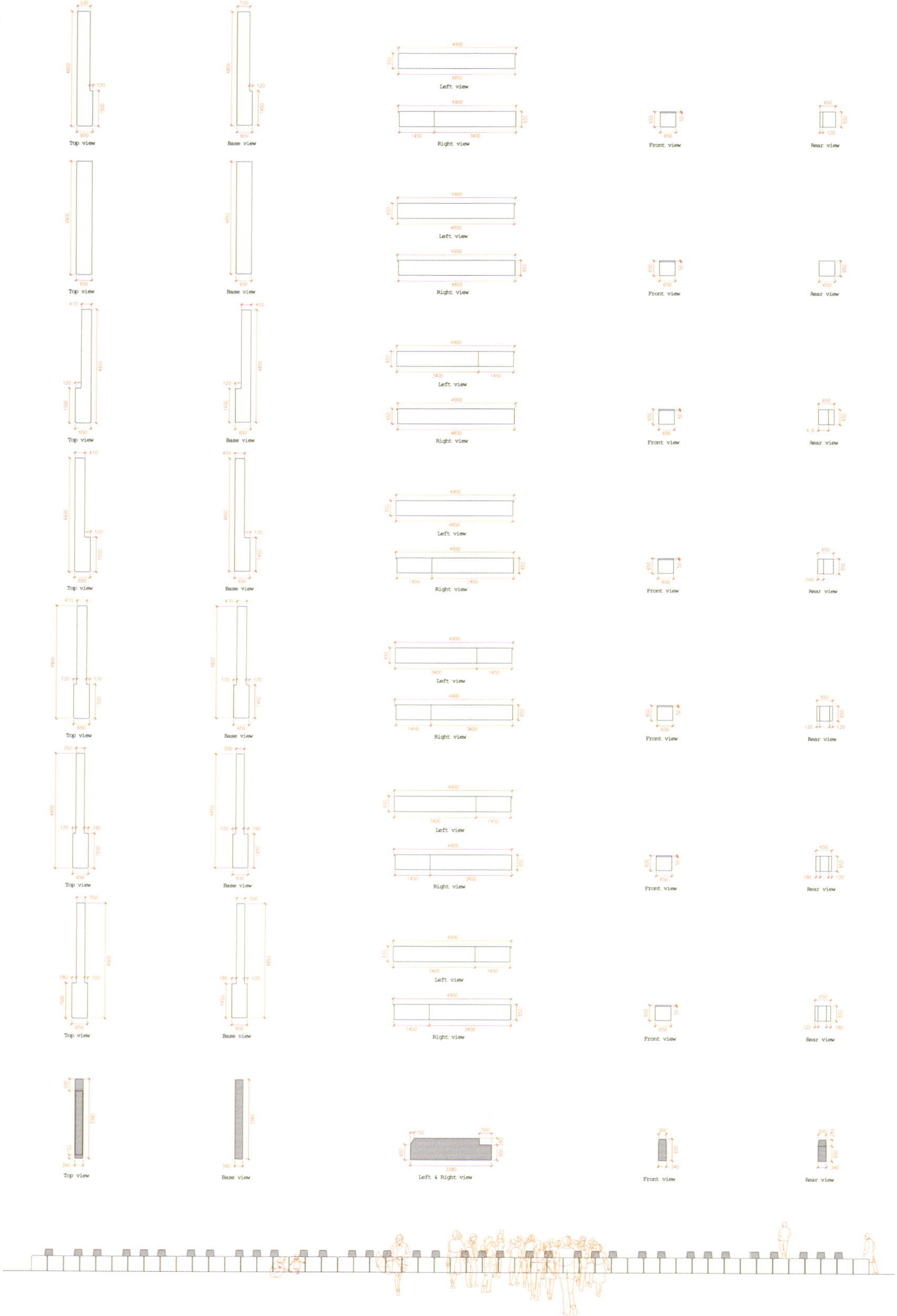
Top view
Base view
Left view
Right view
Front view
Rear view
Top view
Base view
Left view
Right view
Front view
Rear view
Top view
Base view
Left view
Right view
Front view
Rear view
Top view
Base view
Left view
Right view
Front view
Rear view
Top view
Base view
Left view
Right view
Front view
Rear view
Top view
Base view
Left view
Right view
Front view
Rear view
Top view
Base view
Left view
Right view
Front view
Rear view
Top view
Base view
Left & Right view
Front view
Rear view

设计细节

该项目为一个巨型钢琴琴键，不仅美观，且实用性强，适用于社区里各种人群，不论性别、年龄、国籍和信仰。这一建筑通过肖邦后期最伟大的作品之一为设计起到微妙的作用。琴键上的 F1, F2, B4, D5*, G5* 都被按下，这是威尼斯的船歌 Fis-dur op.60 的前奏。这一设计细节为琴键本身赋予了活力，也感染了来往的路人。琴键同时也成为行人们的座椅，给人们一个休憩、反思和深思的地方。也可以把这里作为集会的地方，或者把它作为露天钢琴音乐会的举办场所。因此琴键会成为表演中的一部分，并且是与观众紧密接触的一部分。

Las Negras Waterfront
拉斯维加斯拉斯海滨

Project Location: Almeria, Espana
Architectural Design: Jesús Torres García
Project Area: 6,800 sq.m.
Completion Year: 2010
Photography: Jesús Torres García
Building materials: Wood

项目地点：西班牙阿尔梅里亚
建筑设计：Jesús Torres García 建筑事务所
项目面积：6,800 平方米
竣工时间：2010 年
摄　　影：Jesús Torres García 建筑事务所
建筑材料：木材

A Scenery in Almeria　阿尔梅里亚风景区

Project Description

The landscape of this Almeria area has been used as scenery for forty years: from the birth of a genre such as the Spaghetti Western with Sergio Leone taking the lead with The Good, the Bad and the Ugly and A Fistful of Dollars, to modest productions such as How I won the war by Richard Lester. The fact that such an abundance of artists agrees in the choice of this enclave as a backdrop for their creations gives designers an idea of its scenic quality. The isolation and the lack of development of this region have also helped.

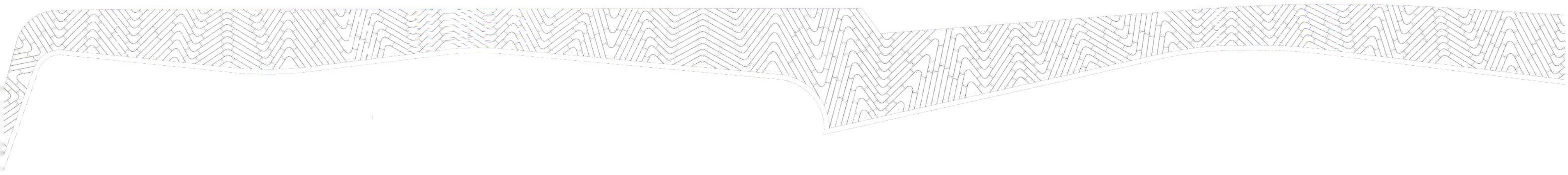

平面图 /Site Plan

Design Details

The example of Las Negras elicits a considerate approach to the public element; the choice of materials has been decisive in consolidating the work within its urban interaction: the wood of the structure and the coverings of perimeter benches encourage a pleasant treatment favouring its consolidation. There is a symmetrical psychology in the human treatment that is here applied to the use and the form of the material as a means of the object's expression. This reflection refers to the natural element, the sound of the sea, the material and shape of natural elements, the vegetation, the geological configurations, as well as the settings of interest. Here the architect becomes a mediator or translator of the values of the context. Possibly, the answer to all these questions would be in a natural conception of the urban landscape, formally spontaneous and amply influenced by its scientific nature: processes, hierarchies and natural materiality.

项目介绍

阿尔梅里亚区已被命名为风景区四十年，像一种流派的诞生，就像从塞尔吉奥•莱奥内导演的"意大利美国西部片如《黄金三镖客》和《荒野大镖客》到适度创作，比由理查德•莱斯特导演的《我如何赢得战争》。事实证明多艺术家认为这片土地给了他们一个非常好的创作背景，这也给该项目的设计创造了灵感。因此这片落后的地也得到了改善。

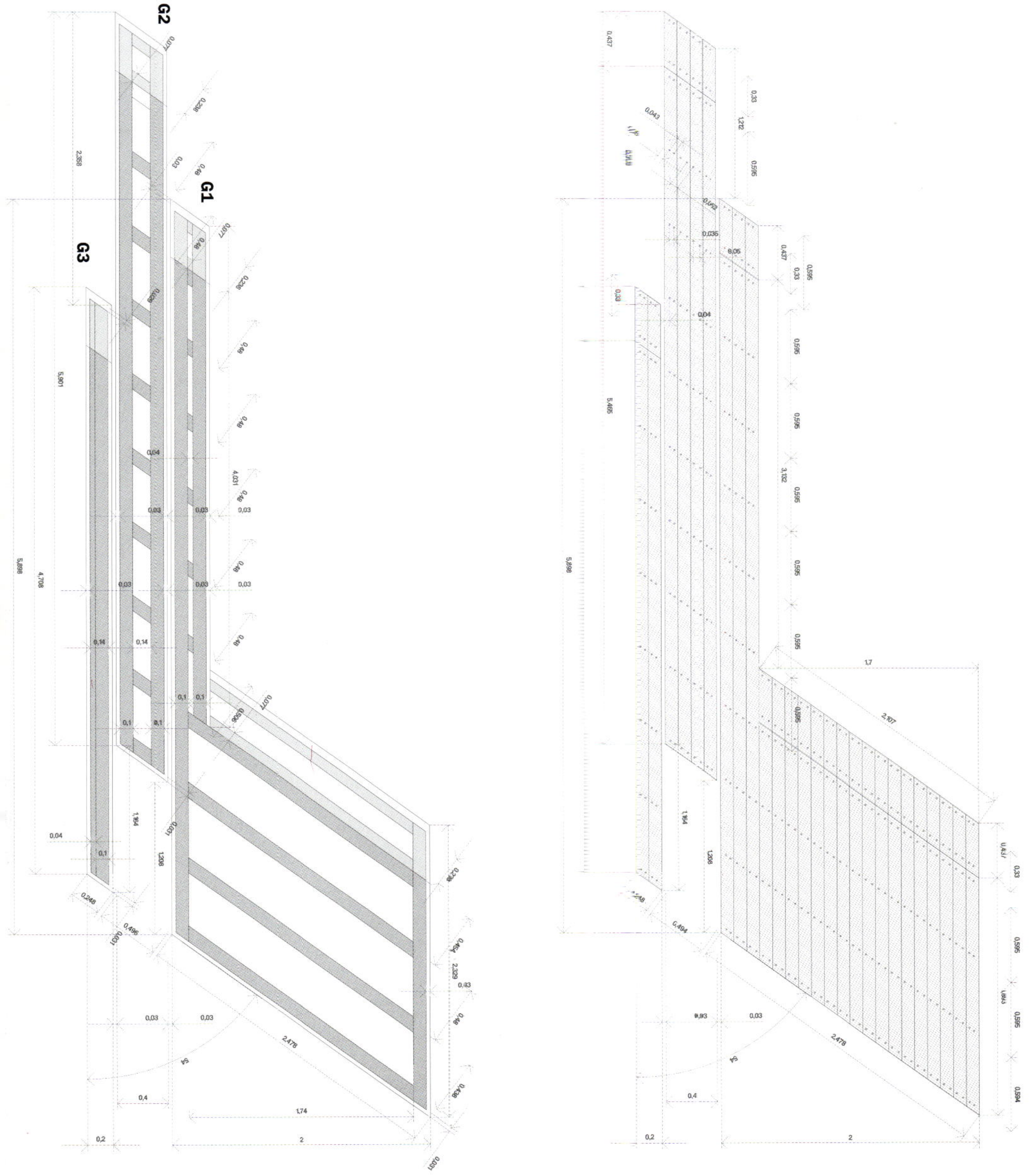

平面图 /Site Elan

设计细节

拉斯内格拉斯滨水区的景观设计包含许多公共元素，设计的进程旨在加强城市元素之间的互动，木头的结构、围边座椅上的覆盖物，都是为了鼓励和巩固公共元素的实施和呈现。在人类医疗上有对称心理这一说法，设计师将其引用进来，材料的选择也显示了物体的表现手法。这种反射设计指自然元素，比如海洋的声音、材料的选择和自然元素的形状、植被、地理结构，以及饶有兴趣的设置。该项目的设计已成为一个中介物或是对背景价值的一种表达。城市自然景观解决了风景区的装饰问题，同时其设计工程、等级和天然材料也受到科学性的影响。

Table Cloth

咖啡桌舞台

Project Location: Los Angeles, USA
Architectural Design: Ball-Nogues Studio
Completion Year: 2010
Photography: Scott Mayora, Benjamin Ball
Building materials: Plywood, Steel Rings, Prefab Table Legs

项目地点：美国洛杉矶
建筑设计：Ball-Nogues Studio
竣工时间：2010 年
摄　　影：Scott Mayoral, Benjamin Ball
建筑材料：夹板、钢环、预制桌腿

A Stage for Performance and Sociality 一个表演与社交的舞台

Project Description

Table Cloth was a temporary performance space in the courtyard of Schoenberg Hall at the UCLA Herb Alpert School of Music in Los Angeles which served as an integrated set piece, backdrop, and seating area for students musical performance and everyday social interaction. It was made of hundreds of individual low, coffee-style tables and three legged stools which linked together collectively to form a "fabric" that hung from the east wall of the courtyard. After the structure served its use as a performance space, the components comprising the installation were dismantled to become smaller scaled commodities, immediately available as coveted products-tables and chairs.

Design Details

The processes of designing manufacturing, assembling, and dismantling the performance space are examples of a unique design and manufacturing methodology that moves beyond and constructive critiques the three "R's" of sustainability-recycling, reuse, and repurposing; processes that typically down-cycle material into less valuable states. After the structure served its function as a performance space, the components comprising the installation were dismantled to become smaller scaled household commodities-tables and seating. This process, referred to as "Cross Manufacturing" by Bell-Hogues, is an integrated design and manufacturing strategy that harnesses digital computation and fabrication technologies to make architectural scaled installations that become collections of smaller scaled products. The items were immediately available and given away as consumer goods, once the installation was dismantled. This approach moves beyond recycling and reuse.

项目介绍

咖啡桌舞台是一个当代表演场所，位于洛杉矶加州大学赫伯埃尔博特音乐学院的勋伯格音乐大楼的院子里。它作为学生表演及社交的场所，有成百上千张低矮的咖啡桌和三张有腿的凳子相连形成一块“布”，挂在院子里东面的墙上。咖啡桌舞台从二楼缓缓延续至地面，散落在地面上的咖啡桌是真正可以供观众休息使用的。由于设计和制作过程中没有破坏咖啡桌的结构，所以当舞台装置被拆除以后，咖啡桌仍然可以继续使用。在这样一种纯粹“材质”的连接下，演员与观众、主动与被动的关系也随之融合。

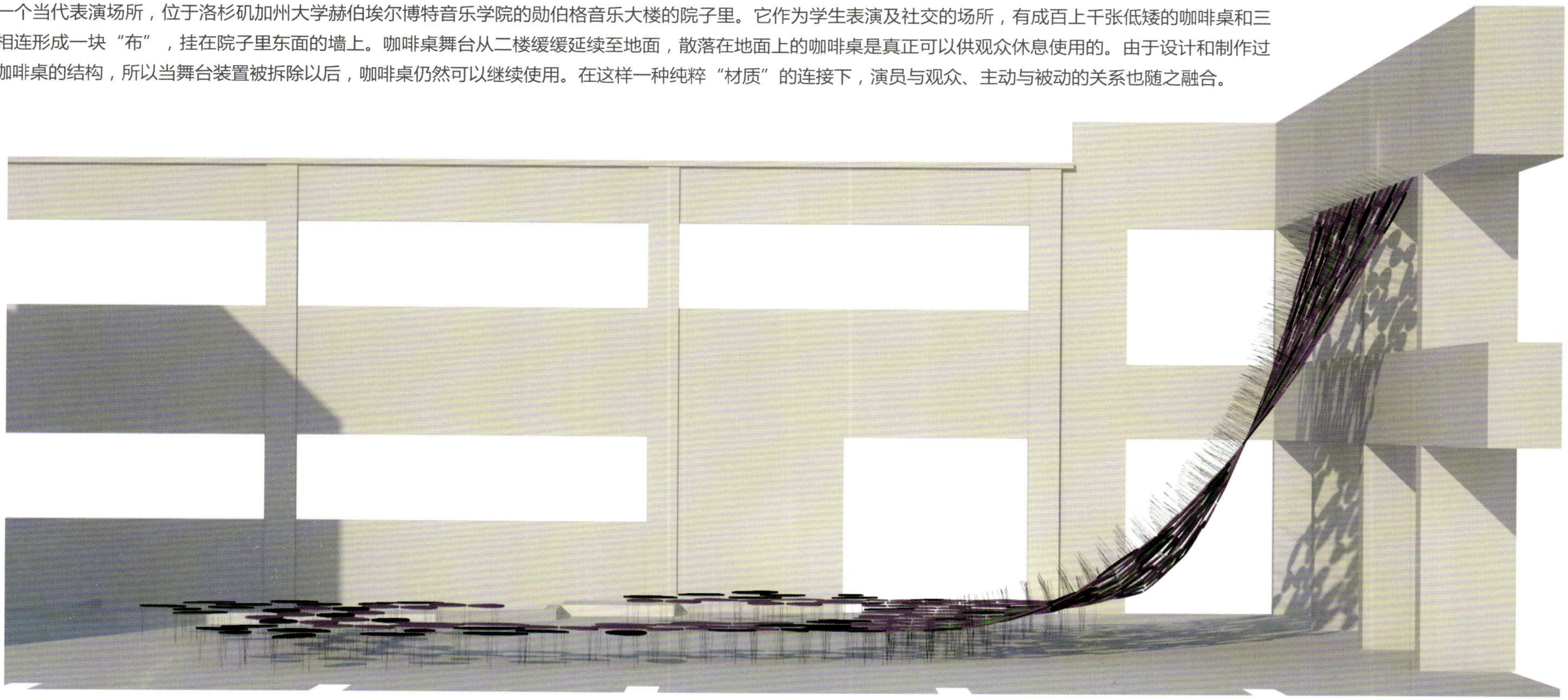

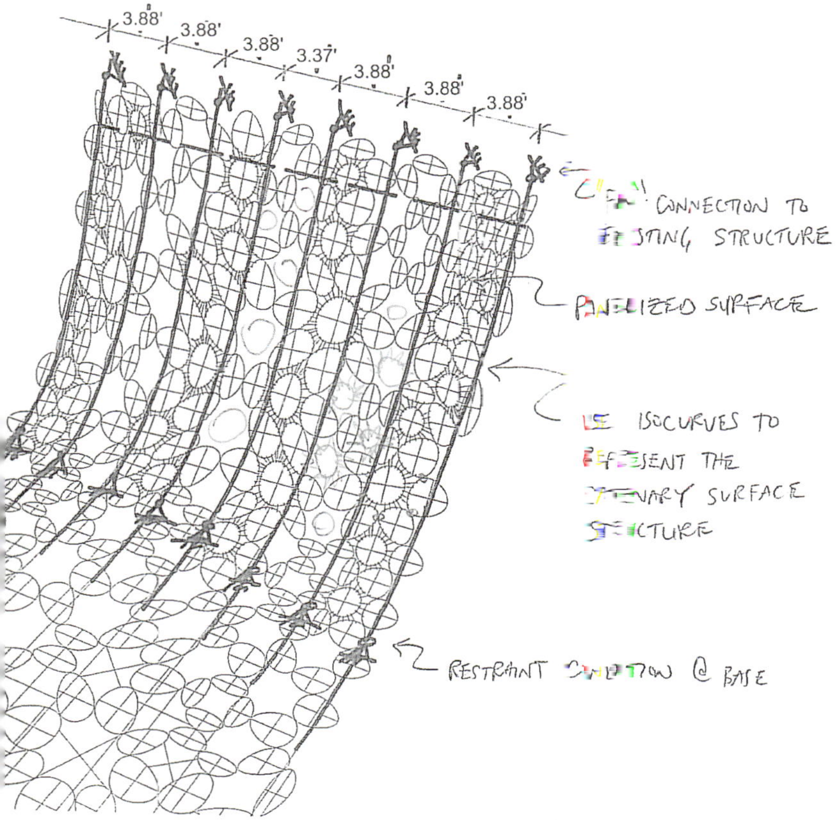

3.88'
3.88'
3.88'
3.37'
3.88'
3.88'
3.88'
CONNECTION TO
STRUCTURE
SURFACE
ISOCURVES TO
THE
SURFACE
RESTRAINT
@ BASE

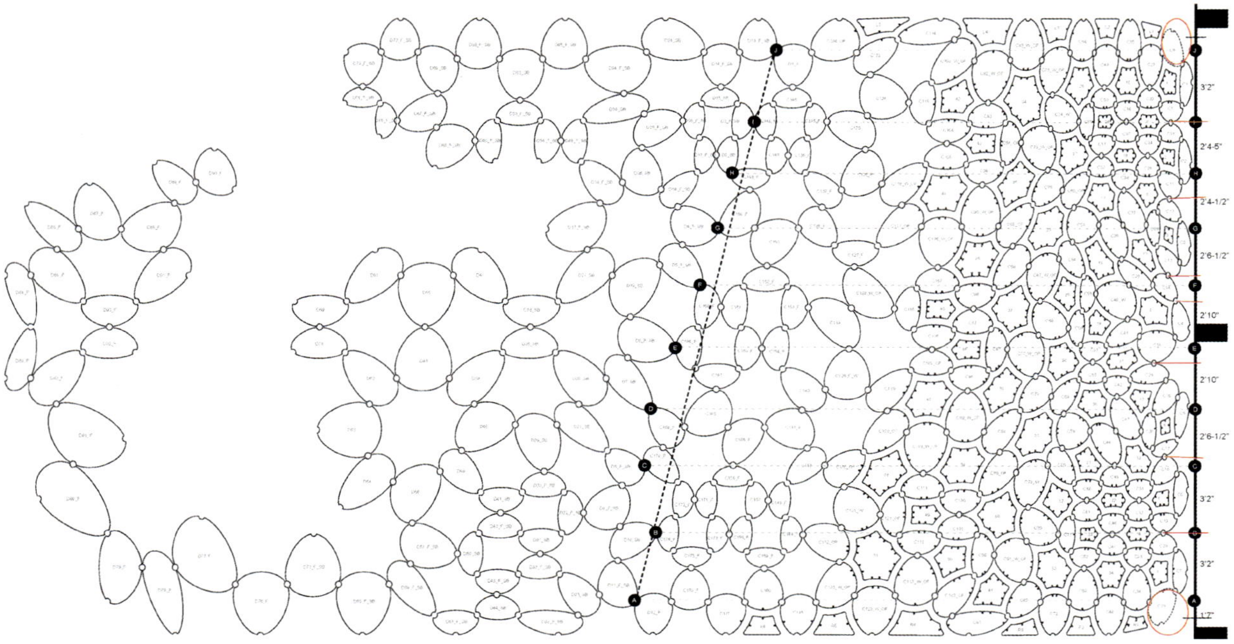

平面图 /Site Plan

Construction Key

C	Upper Curve
D	Flat section
_F	Felt
_W	Warning Sign
_SB	Sand Bag
_OF	Offset Adhesive Felt
SB	Leg at point where surface becomes flat
------	Line to mark end of sloped surface

Distance from the front of the column to the front line of chair legs that need sand bags

A to A	18'6
B to B	17'7
C to C	18'2
D to D	17'10
E to E	16'9
F to F	15'7-1/2
G to G	14'10
H to H	14'2
I to I	13'2
J to J	12'3

设计细节

该项目的创作是一个独特的设计与制造的例子，超越了建设性意见，可持续性三‘R”原理：循环、利用和定位，将具有代表性的下降周期材料转变为价值低廉的材料。该项目作为表演舞台，其组成部分也采用较小规模的家庭用具系列，例如桌子和椅子。这一设计过程正如 Ball-Nogues 所说的，像是“跨界设计”，即一个利用数字计算和制造技术来完成的建筑，也成为小规模产品的集合体。项目所采用的材料都是快速消费品，即使要拆除这一装置，也能轻易做到，这种设计方法超越了循环利用。

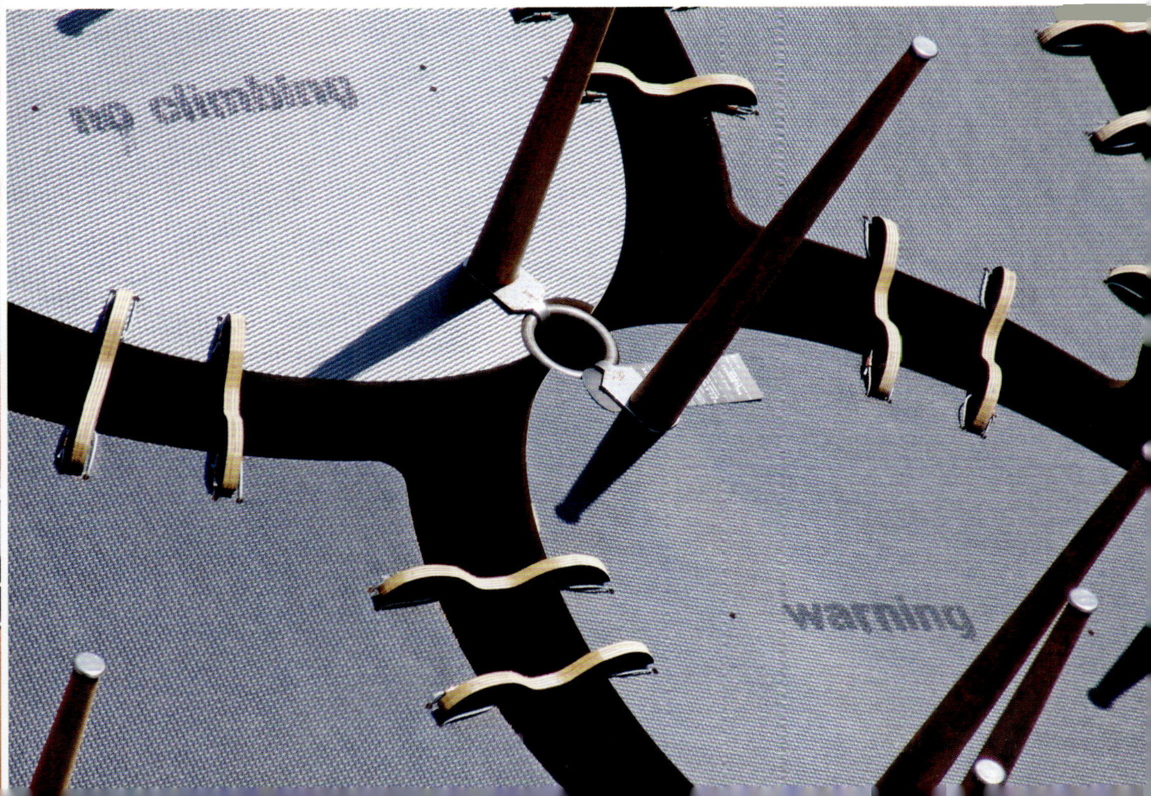

The Umbrella

伞形建筑

Project Location: Mexico
Architectural Design: Javier Senosiain
Completion Year: 2007
Building materials: Ferro-Cement

项目地点：墨西哥
建筑设计：贾维尔·赛诺西设计公司
竣工时间：2007 年
建筑材料：钢丝网水泥

A Place for Sociality 社交之地

Project Description

In the gardens of the Laboratories plant an open area was selected to hold events and banquets for the groups of doctors that visit such plant. This design needed to be integrated to the garden area, to the water mirror and to the exuberant trees without altering the aesthetics of the ecological architecture.

Design Details

That is why this structure, baptized as the umbrella was built, because independently that it meets the objectives-it visually resembles an umbrella corolla, creating with the nervures an undulating area, open at the center to give way to the trunk of a huge laurel of India. It is a sinuous ferro-cement structure of iridescent colors, sustained by a dozen of slim columns that remind us the Gothic style. The cover meets three fundamental purposes: protects from the sun, rain and hundreds of birds that nest in the central tree.

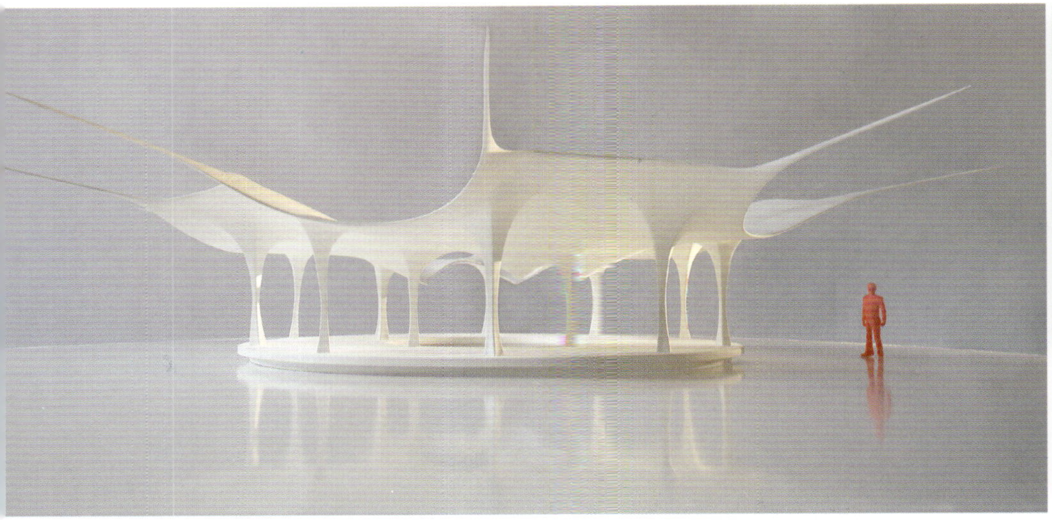

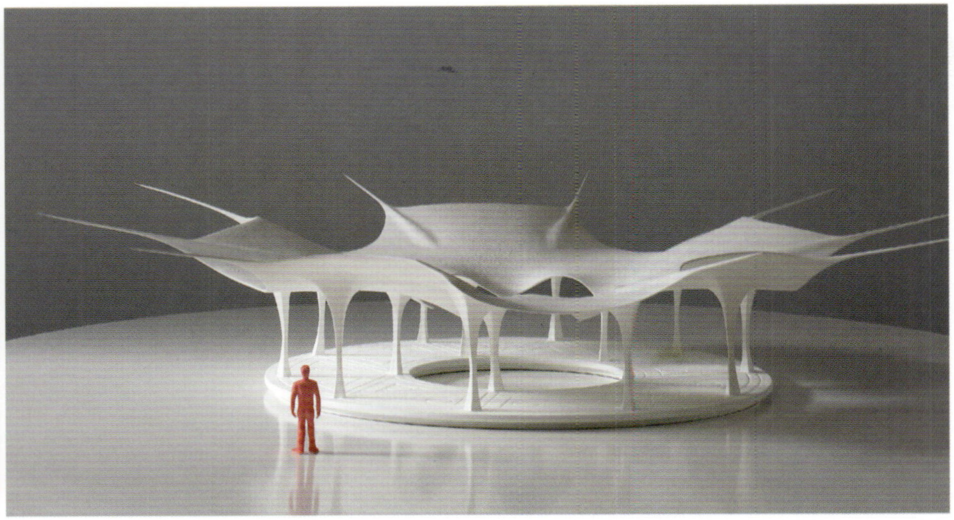

项目介绍

实验室的花园里设置了一个开放区域，为来此参观植物的博士们提供宴会与娱乐场所。整个设计结合了花园区域、水域及生机勃勃的树木，丝毫没有改变生态建筑的美感。

设计细节

整个项目呈伞状，伞脉在视觉上呈现波动感，并在中心张开，为巨大的印度月桂树腾出了生长区域。弯曲的钢丝网水泥结构，五颜六色，而一个个细塔，不禁让人联想到哥特式建筑风格。该项目的外观造型有三个基本用途——蔽光、遮雨和让鸟儿筑巢。

Tierp

蒂耶普

Project Location: Tierp, Swedish
Architectural Design: GORA art & Landscape
Completion Year: 2007

项目地点：瑞典蒂耶普
建筑设计：GORA art & Landscape 设计公司
竣工时间：2007 年

Cone Structures in Mariefred 玛丽弗雷德的锥形建筑

Project Description

With Vägvisaren, the new entrance to Tierp gets a separate and distinctive indication of the flat landscape. The cones are 3.5 meters high and stands in a formation that changes when you move in relation to it. The cones are made of shiny expanded metal that creates a shimmering effect of moire. They are made by hand in Mariefred by the Smith Ulf Haake.

Design Details

Daytime Vägvisaren interact with sunlight. At night, the cones are illuminated from within. The colors are subtle light. They change color very slowly so you do not see shift when you drive past, but when you drive past the next time the color is replaced by another. Pathfinder is as milestones of air and light. They welcome anyone who comes to Tierp.

项目介绍

GORA art & Landscape 设计公司与 Vägvisaren 一起，在通往蒂耶普平缓的地势上，设计了一些建筑物。这些锥体高 3.5 米，矗立成队，稍微移动，便会变换色彩。该锥体建筑位于玛丽弗雷德，是由史密斯乌尔夫哈克用手工制作而成的。

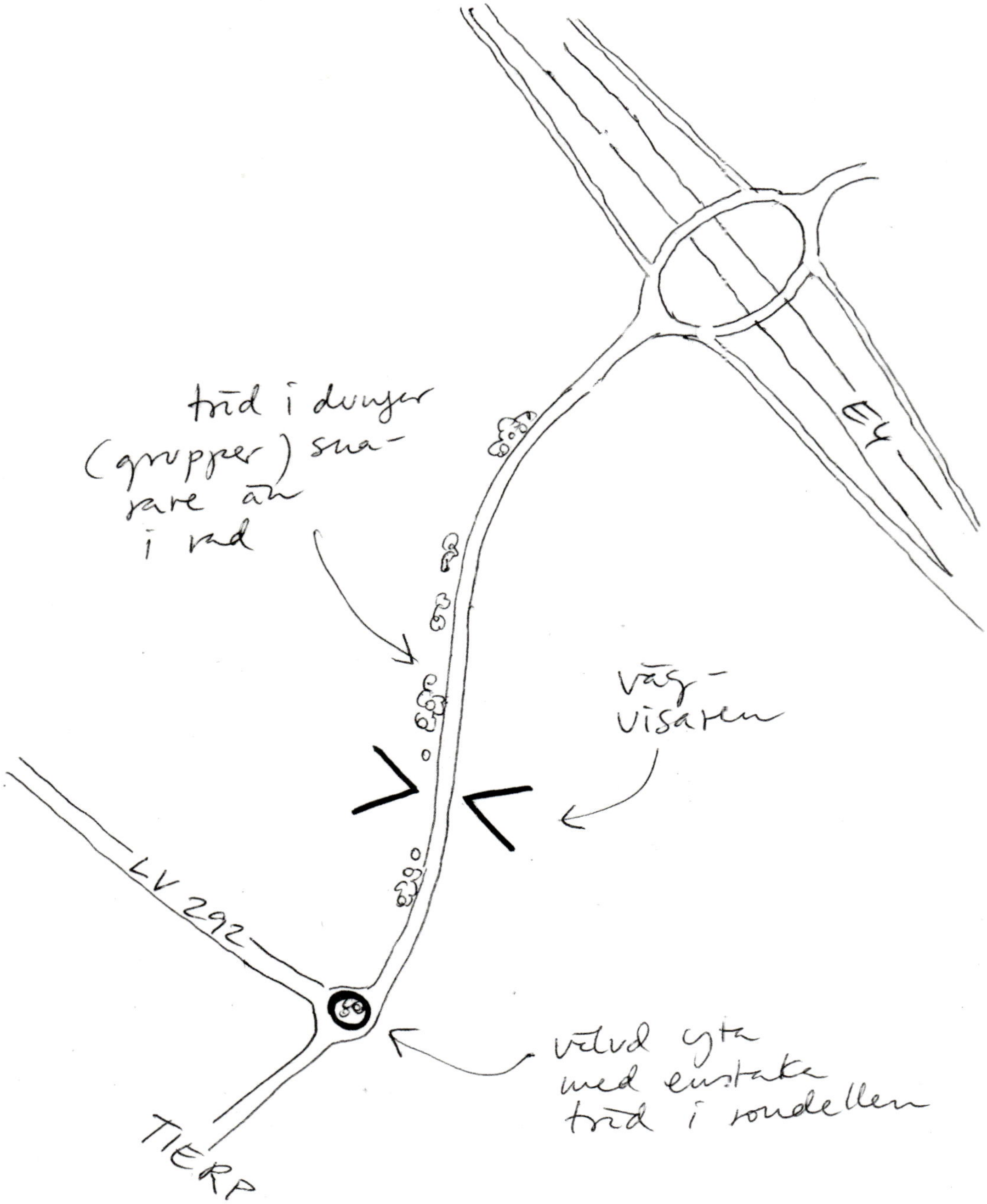

设计细节

白天，Vägvisaren 与阳光互动。夜晚，它便发出光芒，色彩微妙。它们变换色彩的速度非常慢，所以当你驾车驶过，便不会轻易变换色彩，但也许你下次驶过又是另一种色彩。空气和光像里程碑一样作为导航，欢迎着每一个来到蒂耶普的人。

The Conch

“海螺”

Project Location: MapleTree Business City, Singapore
Architectural Design: FARM
Project Area: 500 sq.m.
Completion Year: 2010
Photography: Jeremy San
Building materials: Steel

项目地点：新加坡丰树商业城
建筑设计：FARM 设计公司
项目面积：500 平方米
竣工时间：2010 年
摄　　影：Jeremy San
建筑材料：钢材

An Coral Structure in Maple Tree Business City 丰树商业城的珊瑚建筑

Project Description

Conceived as a part of the lush landscape and the Atoll open plaza at the Mapletree Business City, designers imagined a work that gently reminds one of the various coral forms and lulling experiences at the beach. This particular image keeps playing in their heads: An idyllic stroll along the beach, the designers chanced upon a conch, they hold it to their ears and listen to the sounds of the sea.

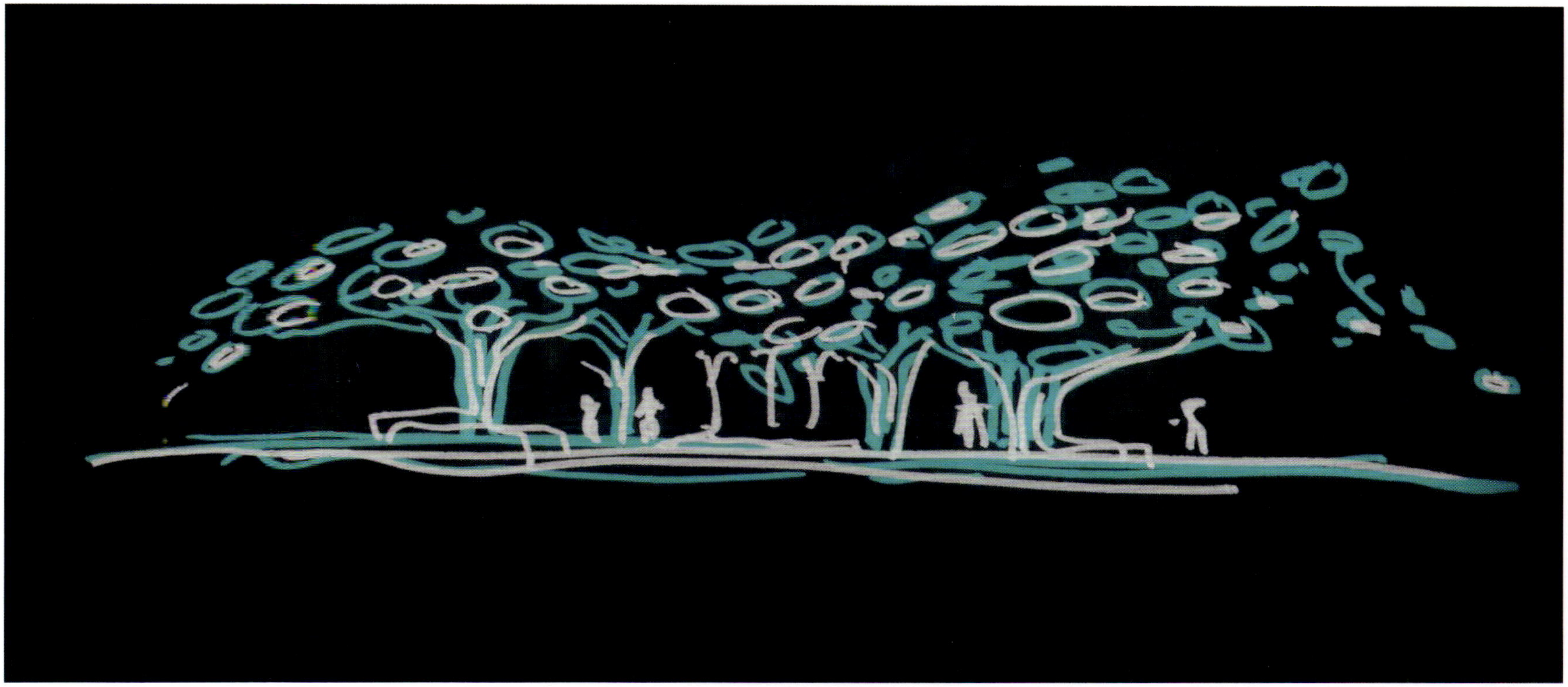

Design Details

"The Conch" is both an outdoor steel sculpture and pavilion sculpted by its environment. Surrounded by tower blocks of the new business development, it presents itself as a highly fluid, wave-like sea of bells-its shape also reminiscent of the trumpet's elegant sprouting form. One can imagine walking beneath the sculpture as if one's below the canopy of a lush forest. And then, from above at the offices' floors, the view is of a singular organic shape land-art form, the multiplicity of parts playing out like the twirling polyps of a rich coral island. It is also a poetic wind instrument, enabling one to listen to the nearby sea, or to the slight movement of the air. Each of The Conch's bells comes together to the ground collectively as stalks. They are dotted with funnels where one can put their ears and listen to the wind. Behaving like nature and taking from nature, "The Conch" is reactive and interacts in its own ways to people.

项目介绍

该项目被认为是郁葱风景和丰树商业城里的环状珊瑚岛的一部分，因此设计师正考虑着有没有一种建筑能隐隐约约暗示珊瑚的各种形式和海滩平静的画面。这个特殊的幻想不断出现在设计师的脑海里。设计师偶尔沿着田园般的海滩漫步，张开耳朵，倾听着大海的声音。

设计细节

“海螺”的设计是根据它的周边环境而设计的一座室外钢雕和亭子。“海螺”四周由新商务中心塔楼环绕，它像海带一样流动着、飘动着，它的形状也让人联想到小号优雅的萌芽形式。可以想象在雕塑下行走就犹如一个人在茂密的森林里行走。从办公楼往下看，“海螺”便成了一个奇特的有机形状的土地艺术形式，大小不一的形状就像是丰富的珊瑚岛上旋转着的水螅。“海螺”也是一种充满诗意的管乐器，能让人静坐，侧听海边浪花，或是感受空气的流动。“海螺”像秸秆一样的钟面慢慢在地面上集合。他们用漏斗做点缀，人们可以贴着耳朵听听风声。“海螺”源于自然，展现自然，并用自己的方式，与人们融为一体，并交相呼应。

The Times Eureka Pavilion

《泰晤士报》Eureka 展亭

Project Location: London, UK
Architectural Design: NEX-Architecture
Completion Year: 2011
Client: The London Times
Building materials: Spruce

项目地点：英国伦敦
建筑设计：NEX-Architecture 建筑事务所
竣工时间：2011 年
开 发 商：《伦敦时报》
建筑材料：云杉木

Bio-Mimicry Design 生物拟态设计

Project Description

The Times commissioned Marcus Barnett Landscape Architects to design and build the garden demonstrating a commitment to science and reflecting the focus of *The Times* monthly science magazine, Eureka. Barnett asked award winning NEX Architecture to design the Eureka Pavilion, and appointed Buro Happold to provide structural engineering. Plant species chosen for the Eureka Garden reflect their benefits to society including medicinal, commercial and industrial uses underlining the fact we could not survive without them. The pavilion design brief was to reflect the same theme.

Design Details

NEX extended the design concepts of the garden by looking closely at the cellular structure of plants and their processes of growth to inform the design's development. The final structure was designed by using computer algorithms that mimic natural growth and is intended to allow visitors to experience the patterns of biological structure at an unfamiliar scale. The primary structure is timber sourced from sustainable spruce forests with a glass panelled roof. The design development of the pavilion focused on the "bio-mimicry" of leaf capillaries being embedded in the walls. The structural geometry was finalised to use primary timber capillaries (300dp x 140wd) to form the basic shape and supporting structure of the pavilion, inset with secondary timber cassettes that hold the cladding. Following completion of the 3D modelling to meet architectural and structural needs, specialist timber fabricators undertook detailed analysis and digital manufacturing of the structure.

项目介绍

《泰晤士报》委托 Marcus Barnett 景观事务所设计一个花园，要求反映泰晤士报科学月刊对科学的探索和追求，Barnett 而后又请来 NEX 事务所设计花园内的 Eureka 展亭，同时请到 Buro Happold 为其提供结构工程支持。花园里植物品种的选择反映了它们对社会的贡献，包括对医药、商业和工业利用等方面的贡献，表达出一个事实，即我们无法离开它们。而亭子的设计也旨在反映这一主题。

crocus

设计细节

NEX 延伸了花园的设计理念，通过观察植物的细胞结构和他们的成长过程，以确定设计的发展。该项目最终的设计利用计算机算式来模拟自然生长状态，能让参观者体验生物结构的图案。该项目采用可持续使用的云杉木搭建，顶部是一个玻璃屋顶镶板。展亭的设计开发重点是将树叶的毛细血管“嵌”在墙壁上的“生物拟态”。展亭的基本形状如同毛细血管，外围包裹着用再生塑料制作的“细胞”。透过框架可以看见花园的景色。展亭坐落在一个 400 mm 深的云杉木梁架上，木架中间以沙袋来压仓，能够抵挡住向上推的强风。

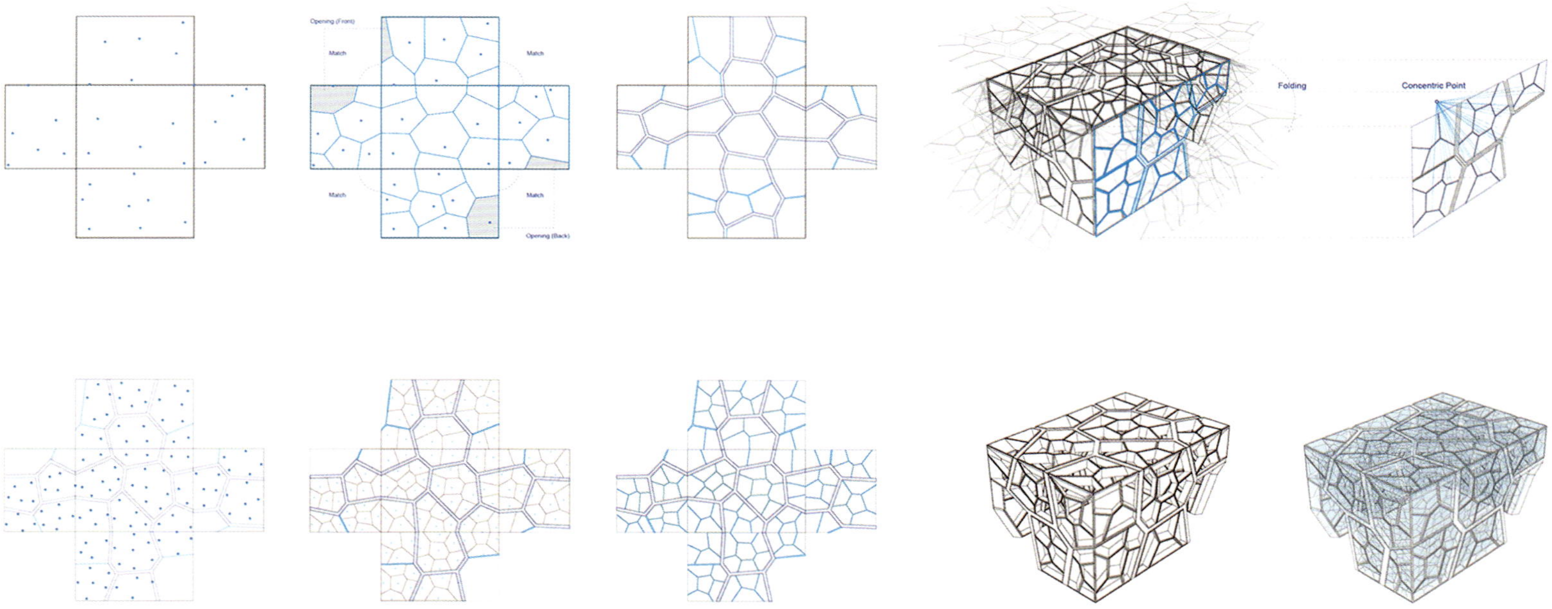

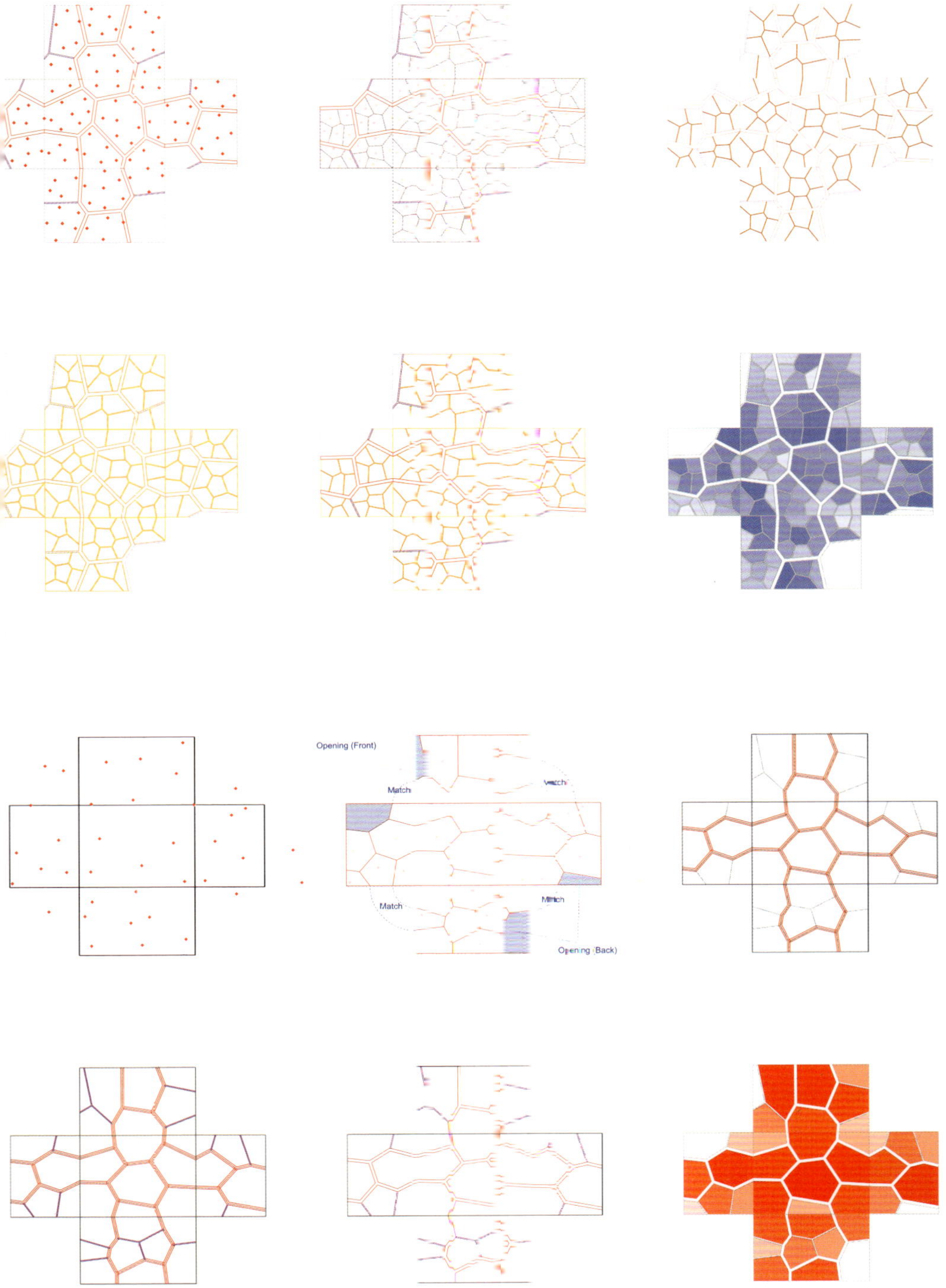
Opening (Front)
Match
Match
Opening (Back)

C.A.A.C

科多巴 CAAC 广场

Project Location: Cordoba, Spain
Architectural Design: Paredes Pino
Project Area: 11,920 sq.m.
Completion Year: 2010
Photography: Paredes Pino & Jorge López Conde
Building materials: Steel, Concrete, Aluminum Plate

项目地点：西班牙科多巴
建筑设计：Paredes Pino 建筑事务所
项目面积：11,920 平方米
竣工时间：2010 年
摄　　影：Paredes Pino & Jorge López Conde
建筑材料：钢材、混凝土、铝板

Colorful Umbrellas Structure 绚烂多姿的太阳伞

Project Description

The design of the plant cover is closely connected to other points of view from which the action is perceived, both existing buildings on the perimeter and from other positions as it happens from the bridge over the tracks of the AVE. Therefore it is of great importance to this treatment of surface color on top. Equally important has been the ground plane treatment as an activator of different uses. The surface treated like a huge board game that can hold any kind of activities, those already known and others yet to be invented. The temporary use of market two days a week is a condition but not in any way a limit, it is only one more of the potential uses to be made.

CAAC USES AND CONFIGURATIONS

MOBILE MARKET SET
168 street vendor kiosk, 168 vehicles

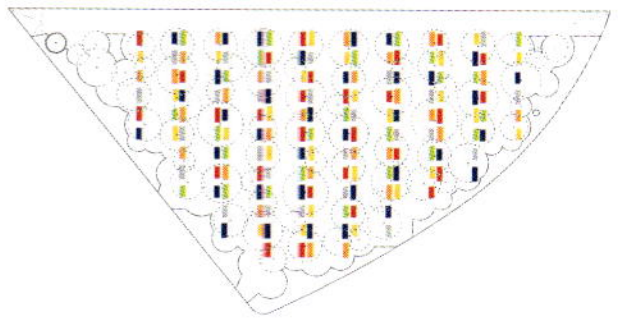

SETTING WITH STAGE

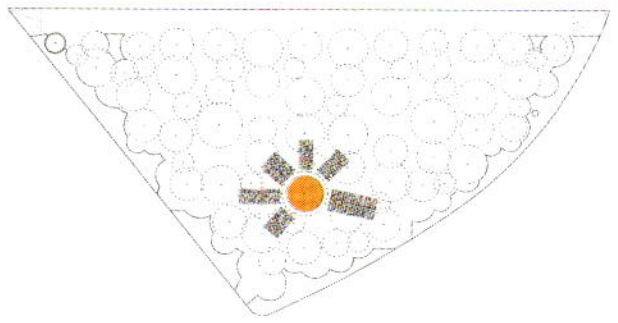

FASHION PARADE SETTING

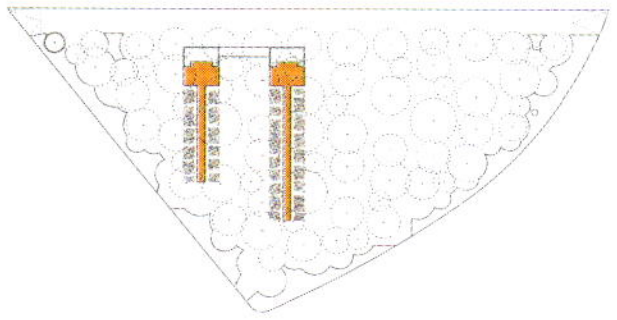

GASTRONOMIC MEETING

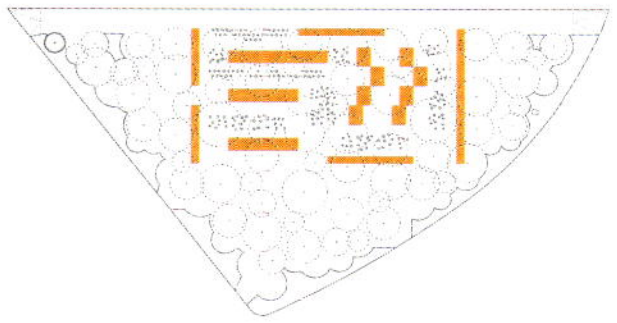

BICYCLE AND KART FACING

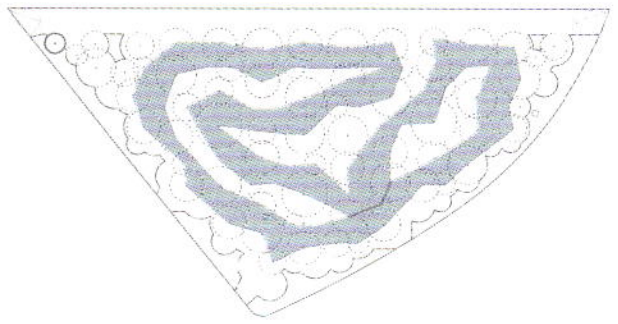

SUMMER OUTDOR CINEMA SETTING

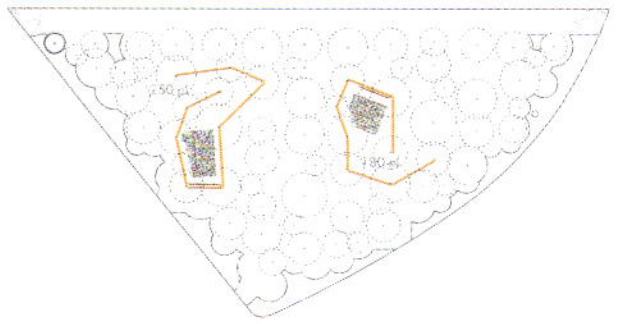

CAAC MOBILITY

MARKET DAY MOBILITY

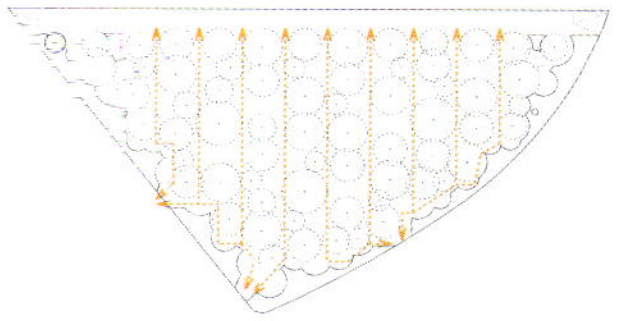

REST OF THE WEEK

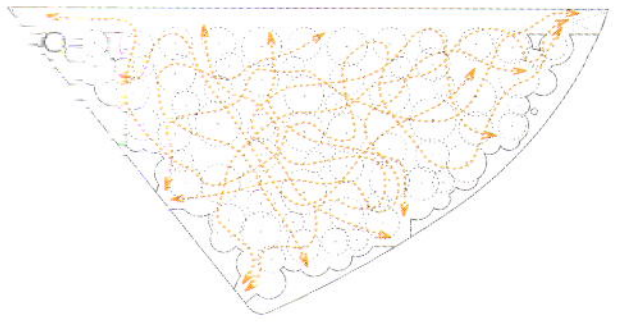

VEHICULAR TRAFFIC

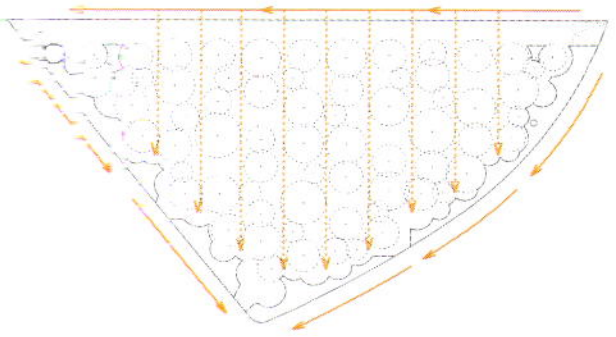

CAAC COMPONENTS OF THE SYSTEM

PETALS UPPER DECK

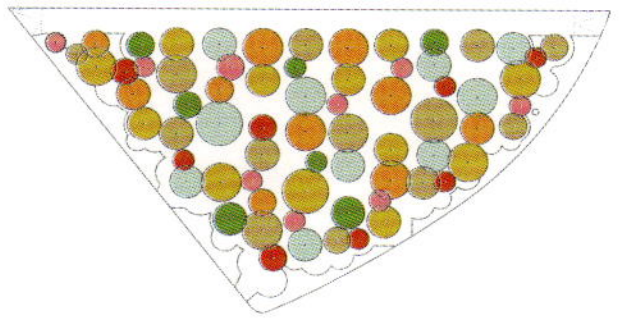

ACTIVABLE PAVEMENT

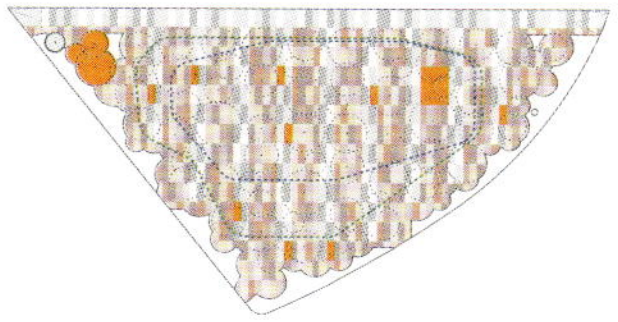

GREEN AREAS

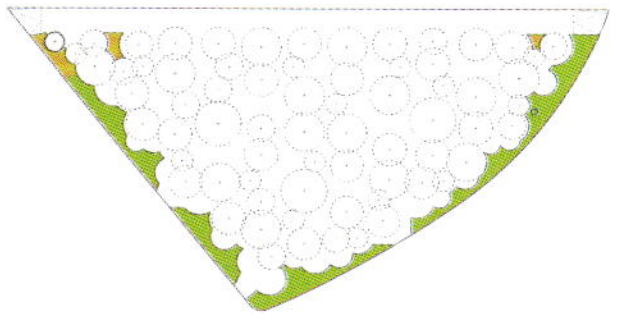

CAAC PREFABRICATION

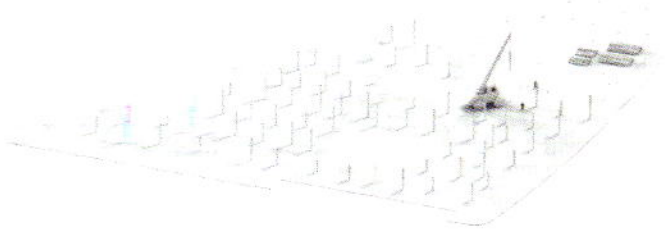

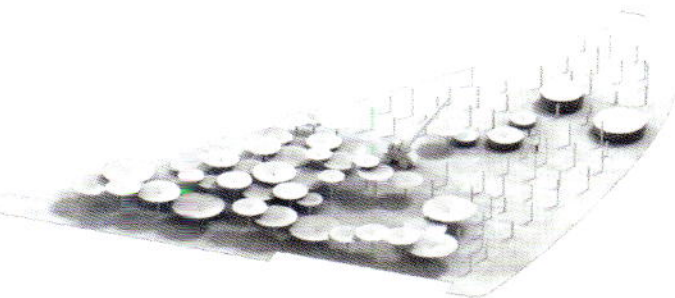

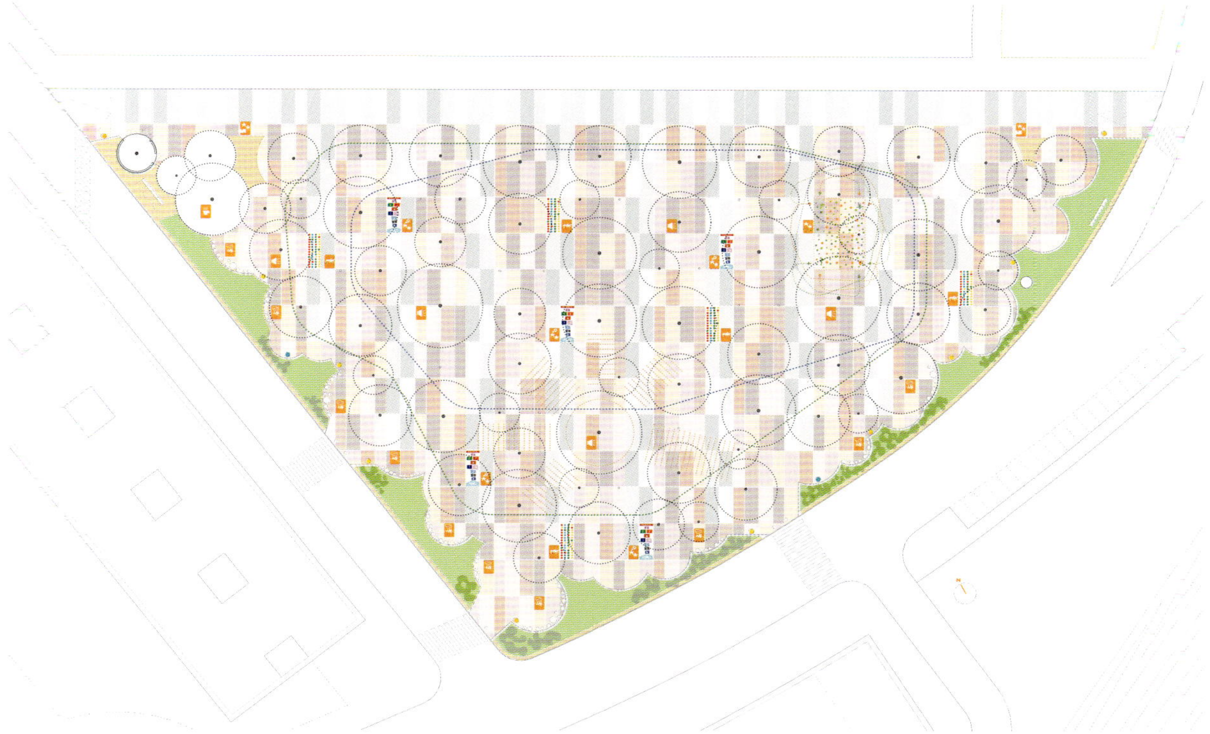

全景平面图 /Panoramic View Plan

Design Details

The cover is solved with parasols diameters between 7 and 15 meters. In turn, the heights are also variable between 4 and 7 meters. This allows great flexibility with little variability and constructive elements. Furthermore, the position of different heights, it avoids excessive opacity of construction, and allows the passage of light reflection through the various umbrellas. The structure is steel umbrellas. The columns inside are constructed with element evacuation of rainwater from the roof, fixed with concrete filling. It is built with aluminum plates in different colors and interior insulation. The treatment of the bottom surface is reflective, giving the place a huge potential in the way that light activates the color by reflection.

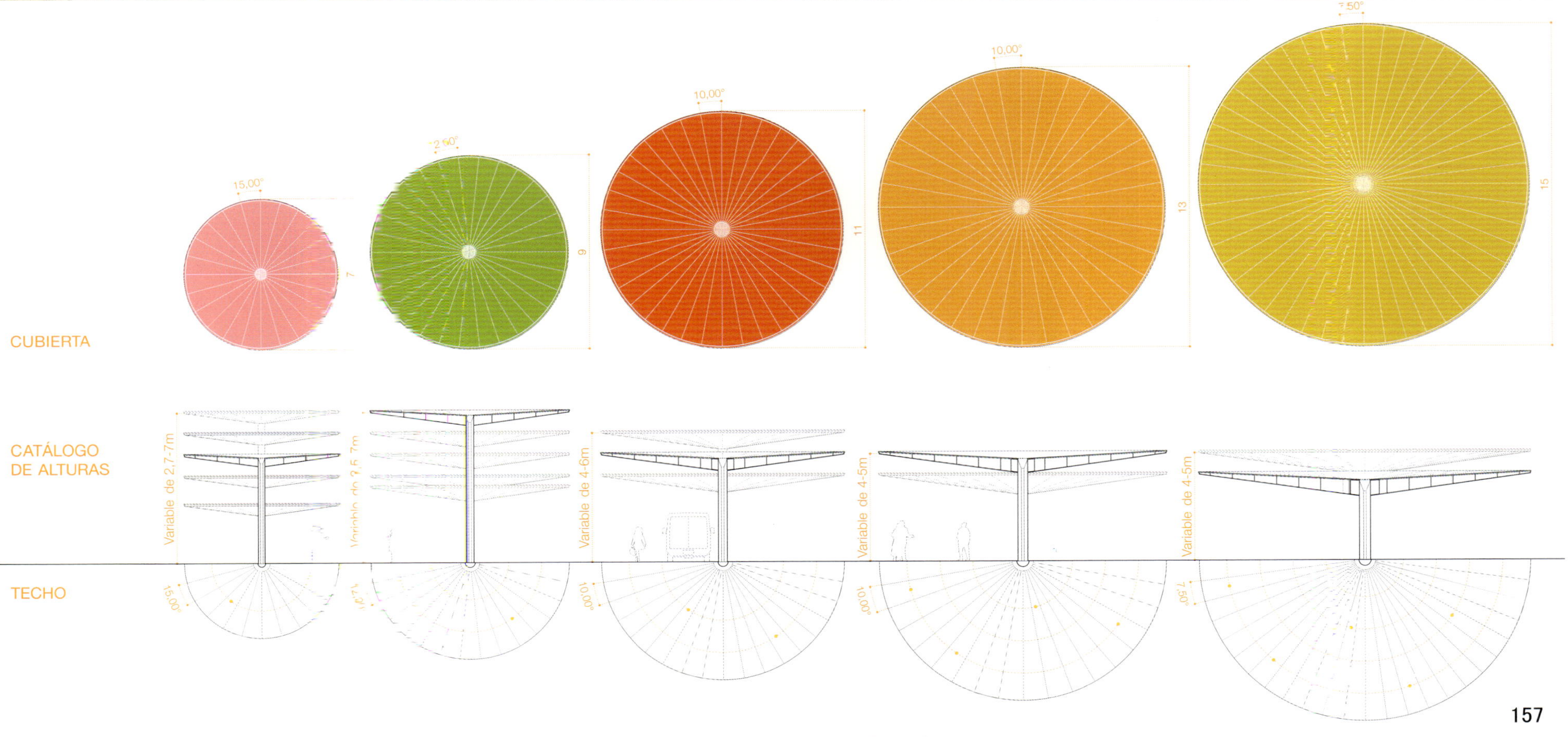

CUBIERTA
CATÁLOGO DE ALTURAS
TECHO
15,00°
10,00°
10,00°
7
9
11
13
15
Variable de 2,7-7m
Variable de 4-6m
Variable de 4-5m
Variable de 4-5m

项目介绍

该项目的设计与它周边的建筑物，及桥边位于高铁轨道上方的其它建筑物紧密相连。因此该项目顶部的颜色设计具有非常大的作用。地平面的设计也具有同等重要的作用，它的设计促进了不同用途的扩展。太阳伞的表面像一个大型的棋盘游戏场，可以用来举行任何活动。太阳伞的设计将使人们不受天气的影响，这将为每周两天的临时市场和其他时间的各类活动提供必要的场所。

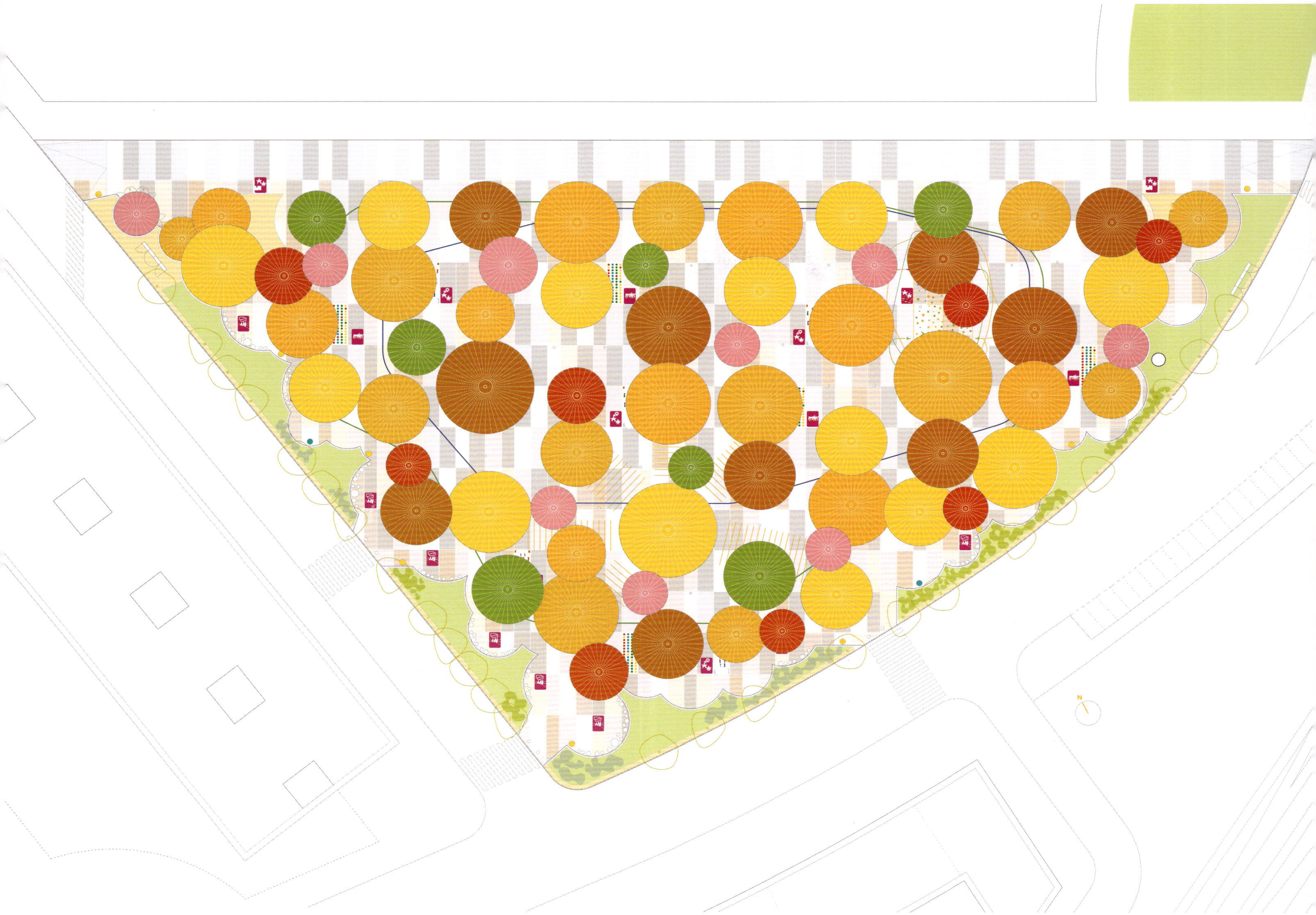

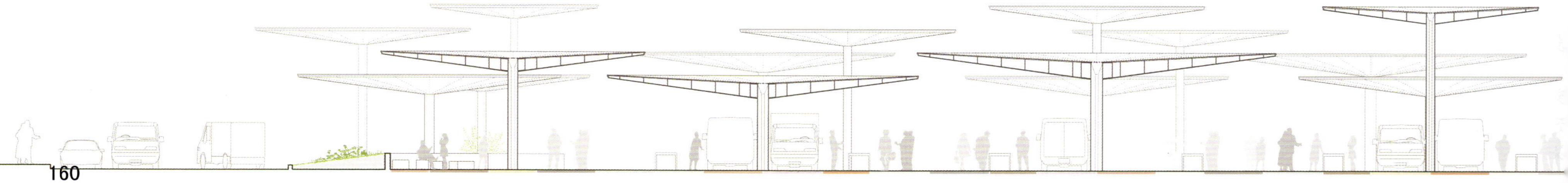

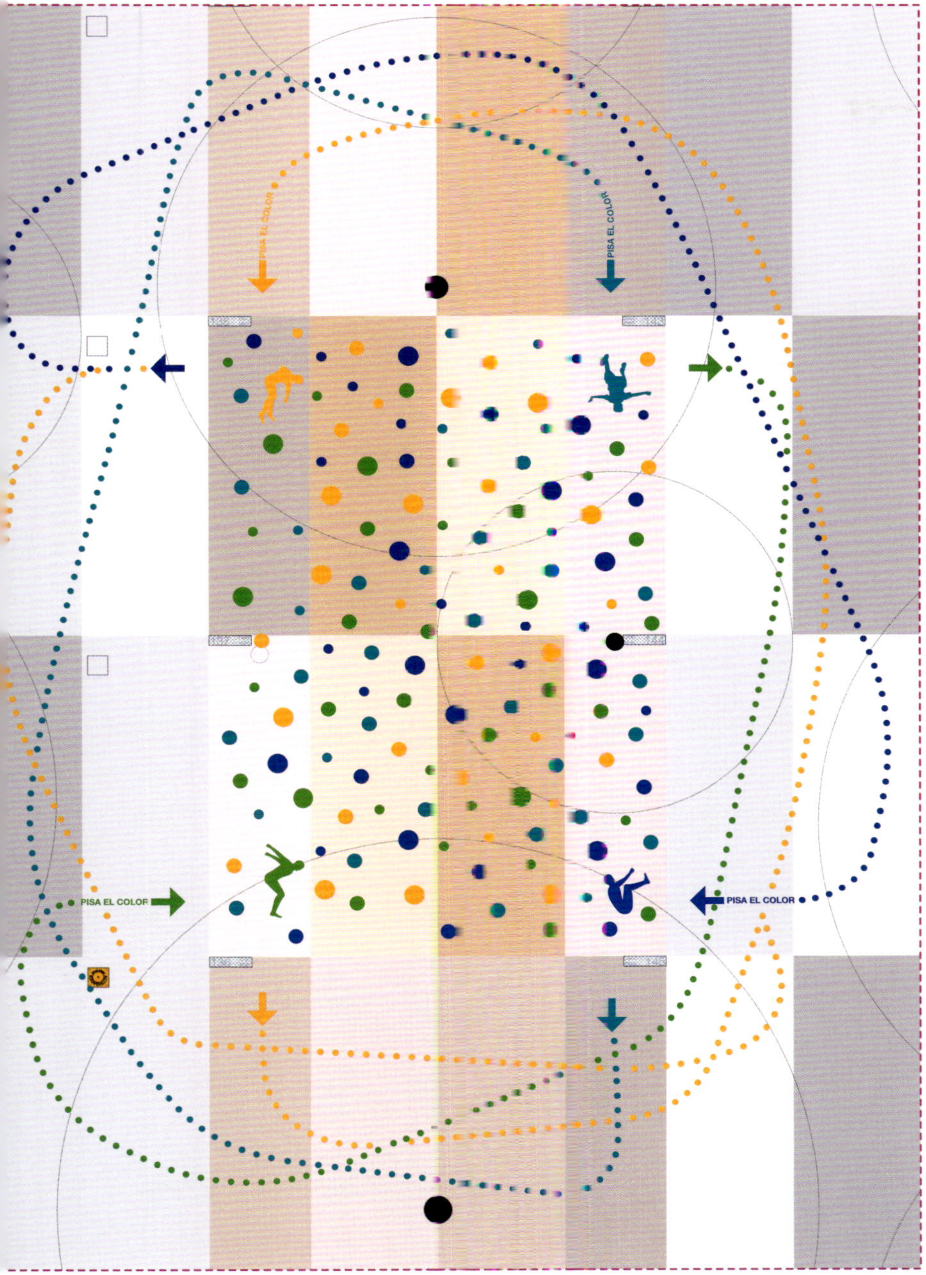

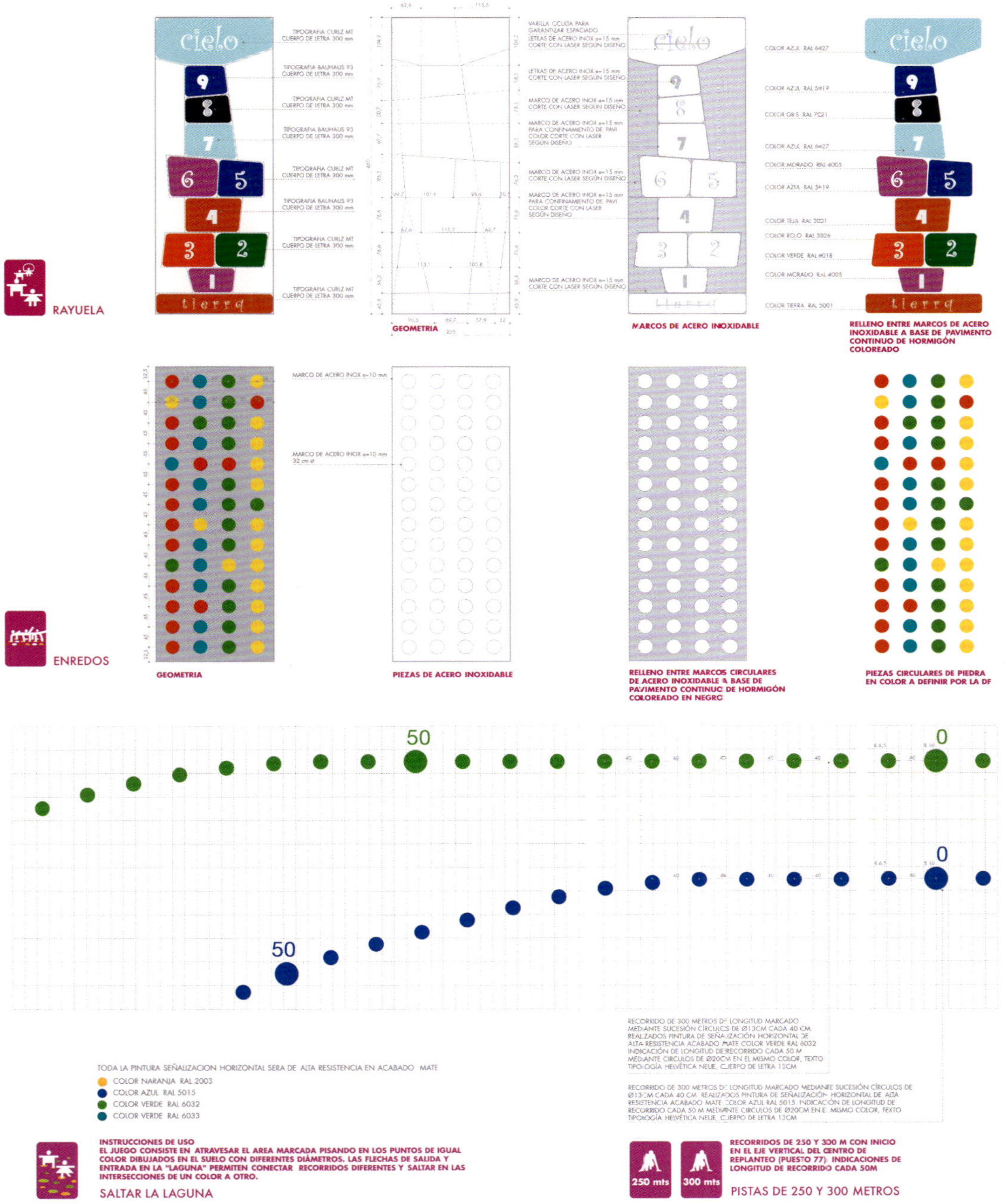

设计细节

该项目的直径为7~15米大小不等。反之，高度在4~7米，主高度和直径都不尽相同的预制圆形建筑组成，灵活随意的排列方式给人一种似曾相识的城市森林下阴影的感觉。此外，不同程度的高度避免了过度不透明性，透过大小不一的太阳伞可引入大量阳光。太阳伞采用钢材制造，伞内圆柱的设计可以疏散顶部的雨水。它用混凝土填充，不同颜色的铝板和内部绝缘设计，也实现了内部排水。太阳伞底面设计成反射面，以一种阳光的反射效果来激活色彩，使该空间具有巨大的吸引力。

Molo

Molo 装饰

Project Location: Slovakia
Architectural Design: Martin Lepej, Michal Riabic
Project Area: 1,000 sq.m.
Completion Year: 2010
Photography: B. Boda, M. Fabian
Building materials: Wood

项目地点：斯洛伐克
建筑设计：Martin Lepej, Michal Riabic
项目面积：1 ，000 平方米
竣工时间：2010 年
摄　　影：B. Boda, M. Fabian
建筑材料：木材

An Open Pavilion in Slovakia 斯洛伐克的开放式展馆

Project Description

The MOLO (pier) project took advantage of a deserted plot right in the centre of the town Martin that was hidden behind a steel fence for years. The authors Michal Riabic and Martin Lepej (Uniform architects) in dialog with town representatives and local entrepreneurs proposed a concept of an open pavilion for this area. Later on, it was the openness that became a central theme for the entire design process.

Design Details

Starting with the matrix of ergonomic parameters, client's expectations were taken into account and 4 sections were defined: stage, entrance zone, bench and auditorium. Stage and auditorium have been equipped with a virtual roof using closed frames. This gives the visitors a feeling of shelter and at the same time provides a bearing structure for the optional canopy roof. MOLO was designed using the Grasshopper plug-in for Rhinoceros 3D. The shape of its wooden ribcage-like structure is defined by 5 longitudinal curves. These served as handles to adjust the pavilion's local cross sections to fit desired functions and to ensure its spatial qualities. This approach has proved very helpful to the overall shape which became a unified fluent morphing of the objects parts.

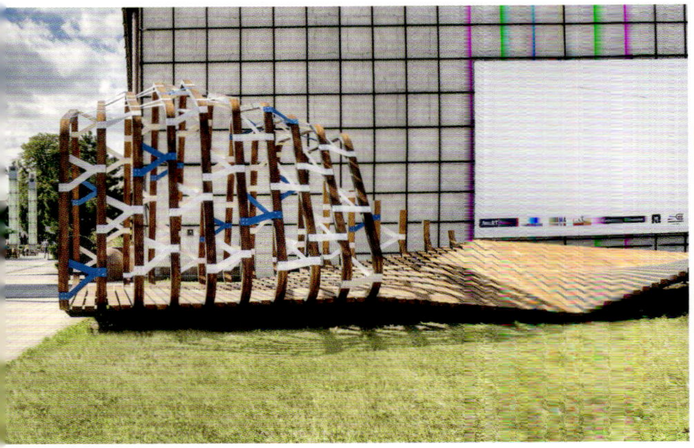

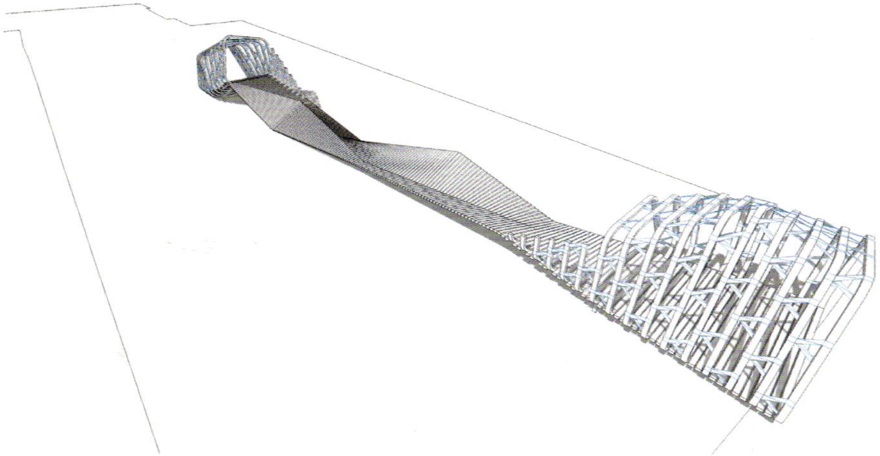

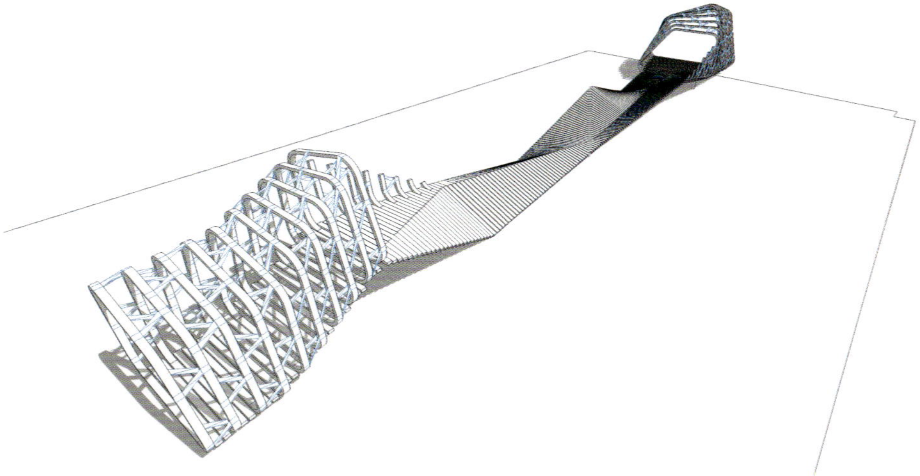

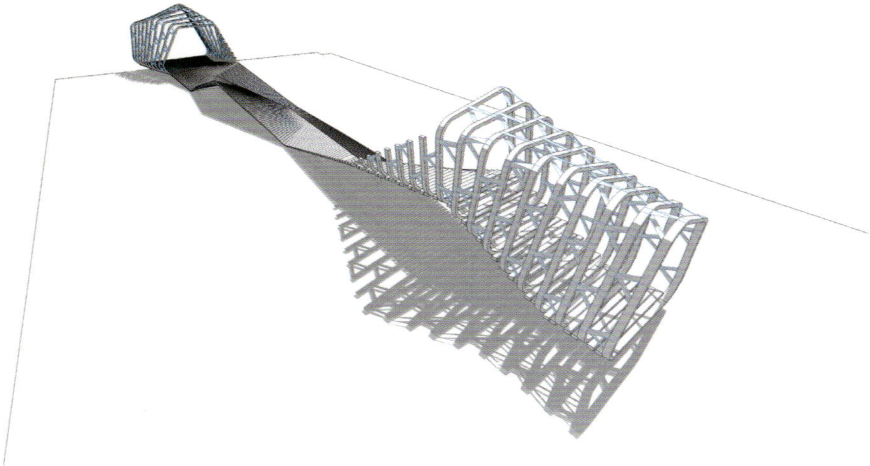

项目介绍

该项目利用市中心里一块荒废的土地，这块地在一排钢护栏旁边隐藏了很多年。设计师们与当地代表和企业商议，希望在这片土地上修建一个开放式展馆。后来，开放式设计的想法也就成了整个设计的核心。

Liquid Space 6.1

透明空间 6.1

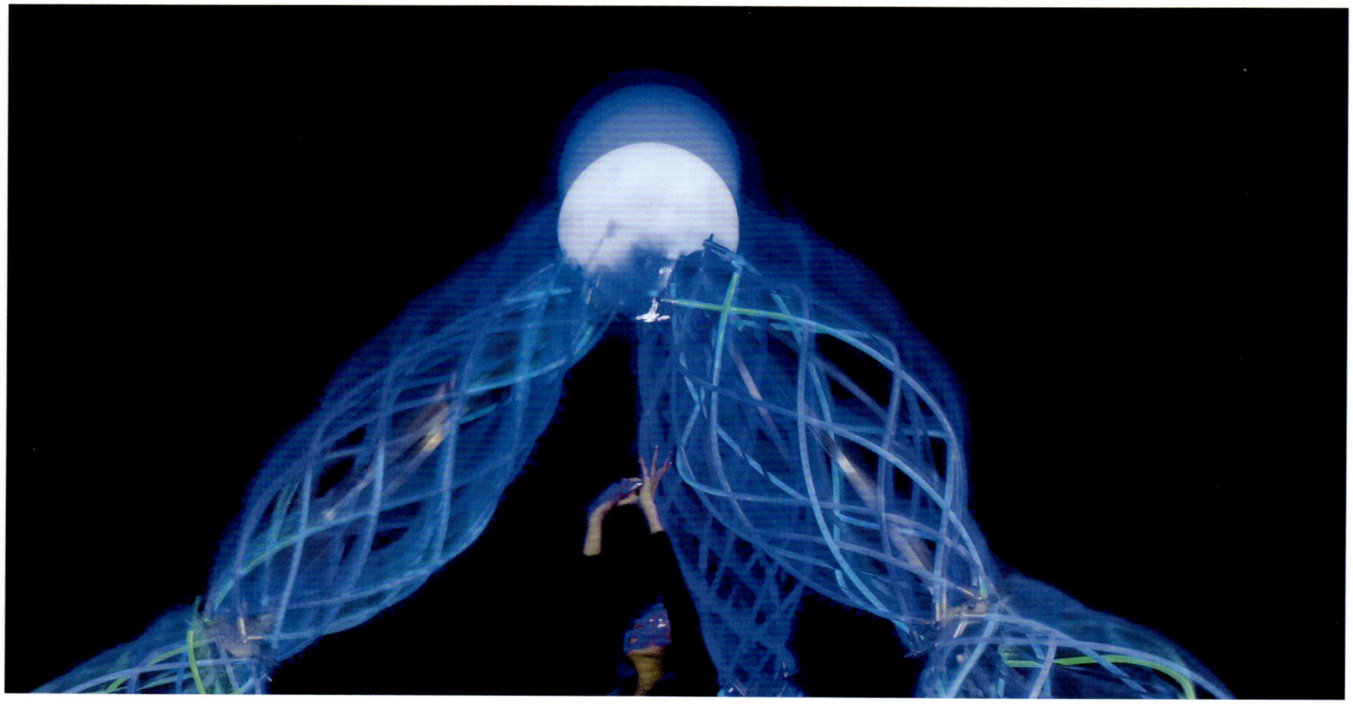

Project Location: Holland
Architectural Design: Daan Roosegaarde
Project Area: 48 sq.m.
Completion Year: 2009
Client: Atelier Oerol & YCAM

项目地点：荷兰
建筑设计：Daan Roosegaarde 艺术家
项目面积：48 平方米
竣工时间：2009 年
开 发 商：Atelier Oerol & YCAM

An interactive artwork 互动艺术装置

Project Description

Liquid Space 6.1 is a public, interactive artwork placed on the island Terschelling NL for Atelier Oerol. Placed in the forest this site related version of Liquid Space further evolved its supernatural interaction with visitors; creating a futuristic merges of humans, technology and landscape.

项目介绍

透明空间 6.1 是一个为 Atelier Oerol 设计的公共互动艺术装置，现摆放在泰尔斯海灵岛上。透明空间 6.1 被置于在森林里，与游客形成超自然的互动，创造出人类、科技和景观的未来形式。

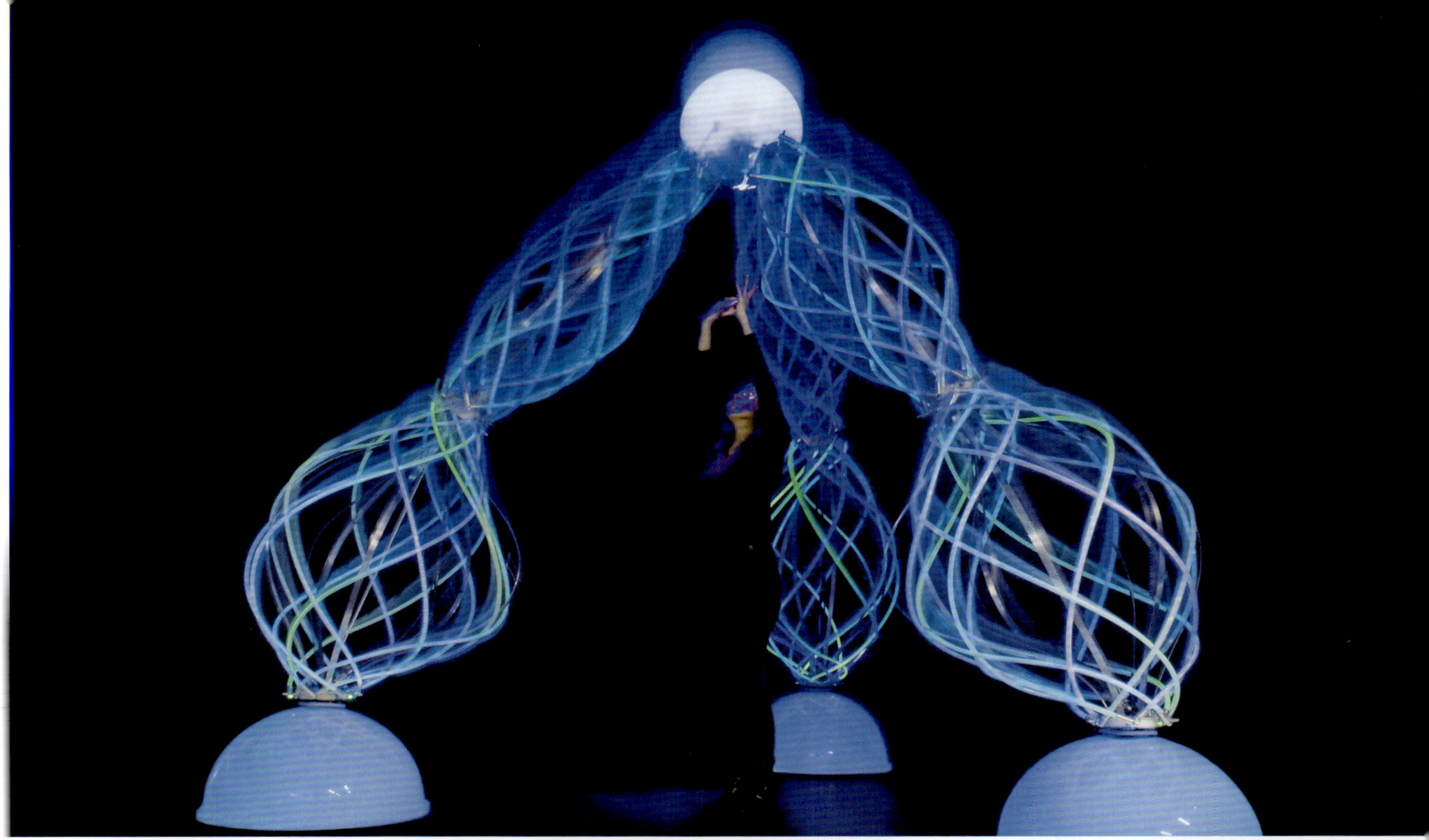

C-Space Pavilion

临时展览馆

Project Location: London, UK
Architectural Design: Architectural Association DRL
Project Area: 100 sq m
Completion Year: 2010
Building materials: Concrete, Steel

项目地点：英国伦敦
建筑设计：建筑协会，AA 建筑研究院
项目面积：100 平方米
竣工时间：2010 年
建筑材料：混凝土、钢筋

A Structure in Bedford Square 贝德福德广场建筑

Project Description

The pavilion was the winning entry to an international design competition held by the Architectural Association School to celebrate the tenth anniversary of the Design Research Lab. The structure, located in Bedford Square in London, was an experimental design research project that was also open for use by the general public.

Design Details

Its visually striking presence invites inspection from a distance and more closely reveals the merging of many discrete concrete and steel elements into a single continuously curved form with a thickened ground that accommodates multiple different uses and modes of occupation. As users moved around the pavilion, its appearance varies from opaque to transparent, producing a complex moir & eacute; that filtered the views of the surrounding park and Georgian terraces. The 12 m span structure encloses a meeting space while also providing a route through for passing pedestrians; blurring the distinction between inside and outside, shelter and stage.

The design and construction of the pavilion made radical use of digital modeling and fabrication techniques to manage over 850 uniquely shaped pieces and 3000 joints of varying angles. Over a period of 6 weeks 16 interactions of the design model were analyzed before a working structural solution was found. In parallel to the digital modeling, numerous rapid prototypes, scale models and full scale physical mock-ups were built to develop the design of individual elements and test the tolerance and fit of entire assemblies. It was finally assembled by a dedicated team of DRL staff and students with assistance from Tecer

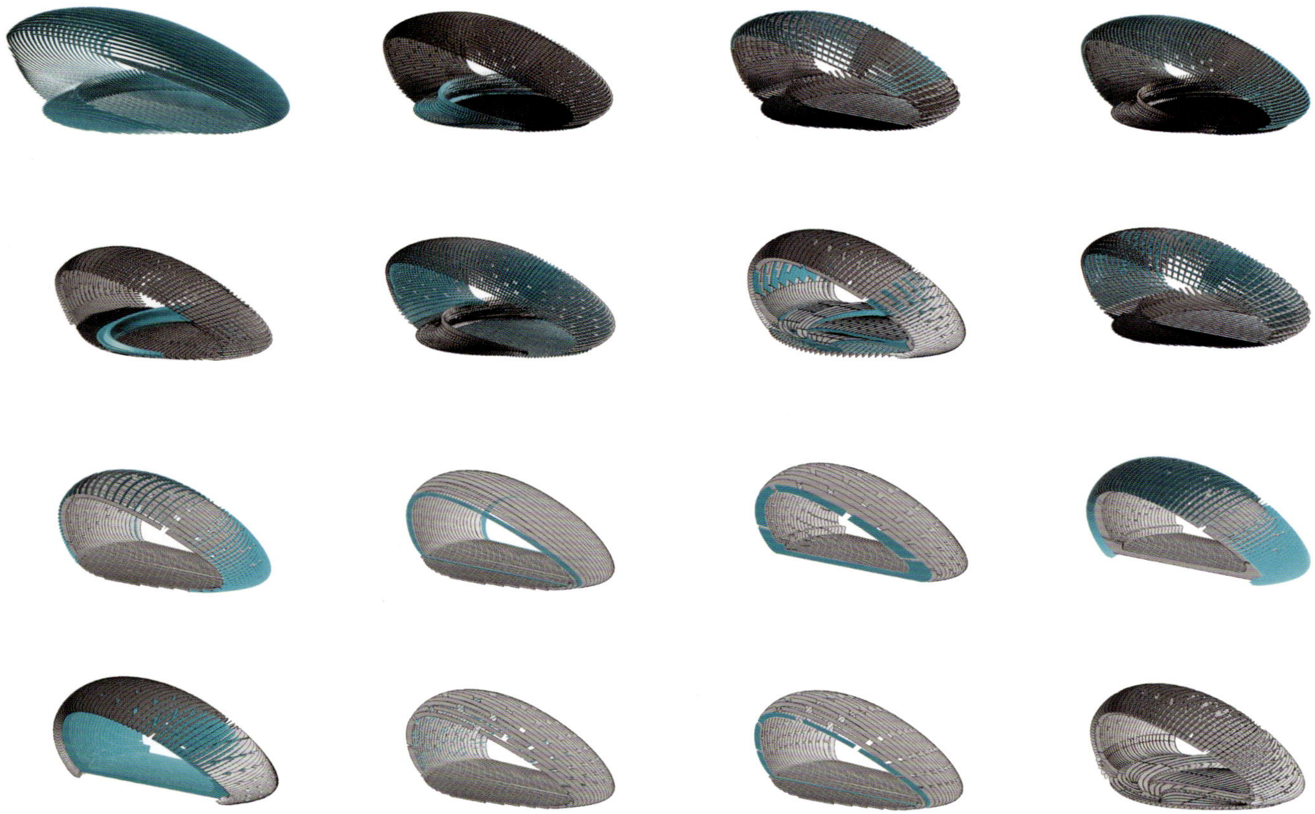

项目介绍

设计临时展览馆是为了庆祝 AA 建筑研究所成立 10 周年，同时该设计也成为了国际设计竞赛上的获奖作品。该项目位于伦敦贝德福德广场，其设计的主要目的是为了进行实验设计研究，但也对大众开放使用。

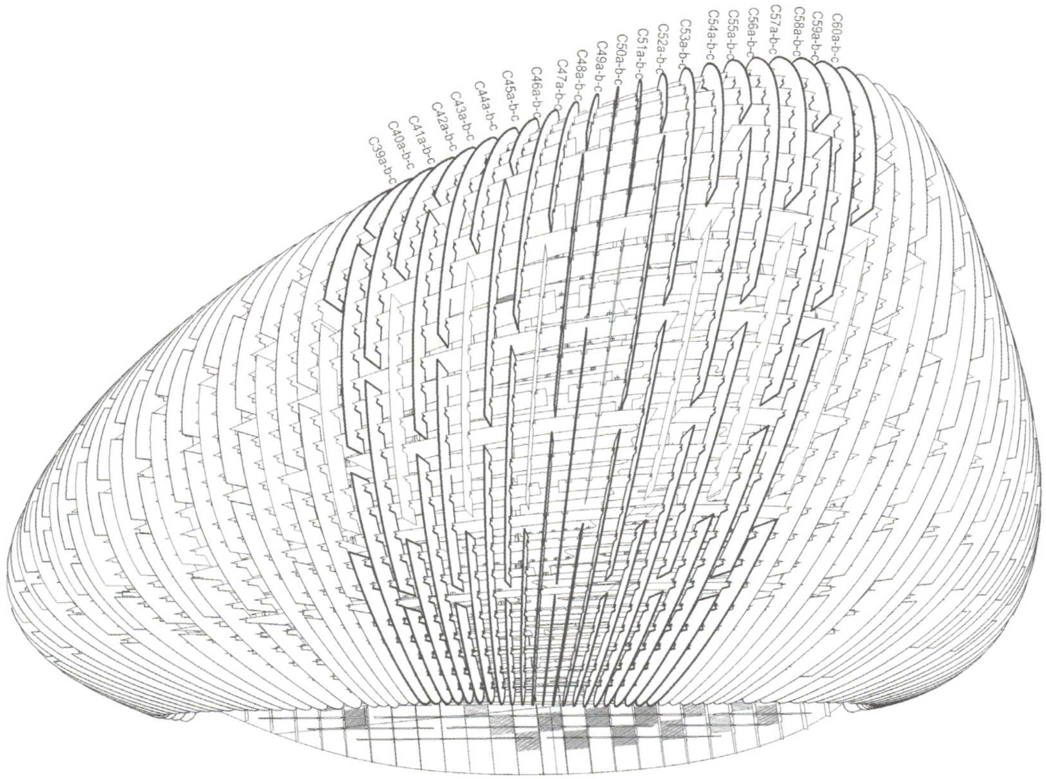

DRAWINGS SHOW INDICATIVE ASSEMBLY SEQUENCE ONLY. REFER TO MAIN CONTRACTORS METHOD STATEMENTS FOR DETAILED CONSTRUCTION PROCEDURES

PROFILES ARE NUMBERED FROM WEST TO EAST
PROFILES ARE LETTERED FROM SOUTH TO NORTH

PIECE NUMBERS:

C39A	STEEL	C39C
C40A	C40B	C40C
C41A	C41B	C41C
C42A	STEEL	C42C
C43A	C43B	C43C
C44A	C44B	C44C
C45A	STEEL	C45C
C46A	C46B	C46C
C47A	C47B	C47C
C48A	STEEL	C48C
C49A	C49B	C49C
C50A	C50B	C50C
C51A	STEEL	C51C
C52A	C52B	C52C
C53A	C53B	C53C
C54A	STEEL	C54C
C55A	C55B	C55C
C56A	C56B	C56C
C57A	STEEL	C57C
C58A	C58B	C58C
C59A	C59B	C59C
C60A	STEEL	C60C

ASSEMBLY SEQUENCE:

PRE DRILL CROSS PROFILE WITH 2NO. 8MM DIA. HOLES FOR EACH GASKET.

REFER TO GASKET ARRANGEMENT DRAWINGS FOR NUMBER AND TYPE OF GASKETS ON EACH PROFILE

INSERT CROSS PROFILES INTO PRIMARY PROFI AND INSERT 2NO. GASKETS INTO EACH JOINT T LOCK IN PLACE

INSERT BOLTS AND TIGHTEN

设计细节

临时展览馆呈现出的视觉冲击让人无论从远处还是近处便能一眼察觉，它的设计反映出许多融为一体的离散混凝土和钢结构，以及连续弯曲且表面增厚的建筑体能适应多种用途。由于使用者在展览馆里的摆放不同，其外观也不尽相同，从不透明到透明，也使得周边花园与乔治亚平台的景色映入眼帘。临时展览馆是一个有着12米跨度的建筑体，包含1个会议室和人行道，这一设计将室内外的界线渐渐淡化。

临时展览馆的设计采用数字化建模和制造技术，将850片样式独特的材料板和3，000多个有不同角度的节点结合起来。在确定结构方案之前，16个设计模型的相互作用让设计师们一直讨论了6个多星期。为了平行数字化建模，许多实验原型、比例模型和满刻度的物理模拟也被运用起来，以开发个别元素和测试公差。最后的组装是在里德的帮助下由DRL特别小组和学生完成的。

Visual Permeability Pavilion

视觉通透休闲凉亭

Project Location: United States
Architectural Design: Steven Sanchez, Aaron Mark, Aaron Ber
Completion Year: 2011
Photography: Warren Berman
Building materials: Timber

项目地点：美国
建筑设计：Steven Sanchez, Aaron Mark, Aaron Berman
竣工时间：2011 年
摄　　影：Warren Berman
建筑材料：木材

A Multiple Spaces in Columbia University

哥伦比亚大学的多功能空间

Project Description

Recently completed as part of a digital fabrication course at Columbia University's Graduate School of Architecture, the Visual Permeability Pavilion was designed and built as part of their end of the year show and graduation ceremonies. The purpose of this pavilion is to provide multiple spaces for relaxation, contemplation, and social interaction.

Design Details

The two spaces within the project are broken apart, providing one space for two people to relax in a more private setting, and one space for four people to have a conversation or drink. The angling of the wooden slats was designed to maximize this separation for the private zone, and minimize it for the public zone; thus creating a gradient of visual permeability. The form was derived from one continuous strip that wraps around it, while touching the ground in minimal locations. The density of the slats is based on three density types. The densest zones are designated for walking and sitting, the medium zones are designated as backrest locations, and the lightest zones are meant for shading. This pavilion will act as a beacon to entice a passerby to come over, sit, and relax.

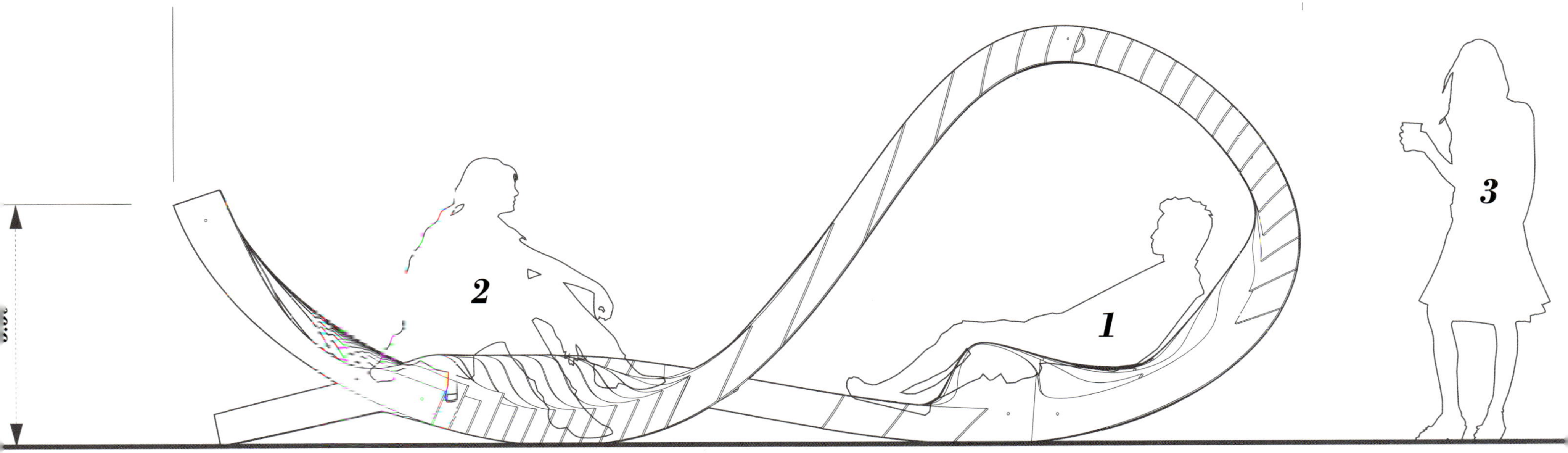

项目介绍

目前在美国哥伦比亚大学建筑学院研究所，一个视觉通透休闲凉亭，作为数字化制造课程的一部分已设计完成。视觉通透休闲凉亭的设计也成为年终展览和毕业典礼上的一部分。该项目设计的目的在于为人们提供一个可供放松、沉思和社会互动的广阔空间。

设计细节

凉亭主要包含 2 个独立的空间，其中一个私密空间可容纳 2 个人同时休息放松，另一个空间可容纳四 4 个人交谈或畅饮。设计师利用木条板的角度划分出这两种区域（私密区角度最大化，公共区角度最小化），因此创造了一种倾斜的视觉通透感。视觉通透休闲凉亭的形状设计来自于一条连续的“带形”，并以最小的点与地面接触。这些条板的密度划分为 3 类：密度最强的地方用来行走和坐卧；中密度区域作为靠背倚靠；密度最小的区域用来遮阳。

Wind Shelter

遮棚随风转

Project Location: Blackpool, UK
Architectural Design: McChesney Architects
Completion Year: 2006
Photography: Peter Cook

项目地点：英国黑池
建筑设计：McChesney 建筑事务所
竣工时间：2006 年
摄　　影：Peter Cook

A Rotating Shelter in Blackpool　英国黑池的旋转挡风建筑

Project Description

The brief was pretty prescriptive. Designers wanted a rotating shelter, driven by a vane pointing into the prevailing wind, in order to provide constant protection from the elements. The only refinements to the original shape were due to the wind. They wanted to use a wind tunnel, but the client would not go for it because of the cost, so they did it themselves. You can do this thing with a fan, and if you have got a bit of string and a pencil in your hand you can see if it works. Designers tried a lot of different shapes with this very simple method. In the end, they came up with one which they suggested for the prototype.

Design Details

Designers couldn't use any compound geometry. There are boat builders who can do that sort of thing, but it was out of their price range. In any case, they were worried that it would look a bit too fussy. It's actually quite simple. There is a central spine, an axis of symmetry up the middle, so you can set up perpendicular sections and cut each one. Above the base plate, the structure has a skin and below it the fins just extend down and then cantilever out to pick up the bench.

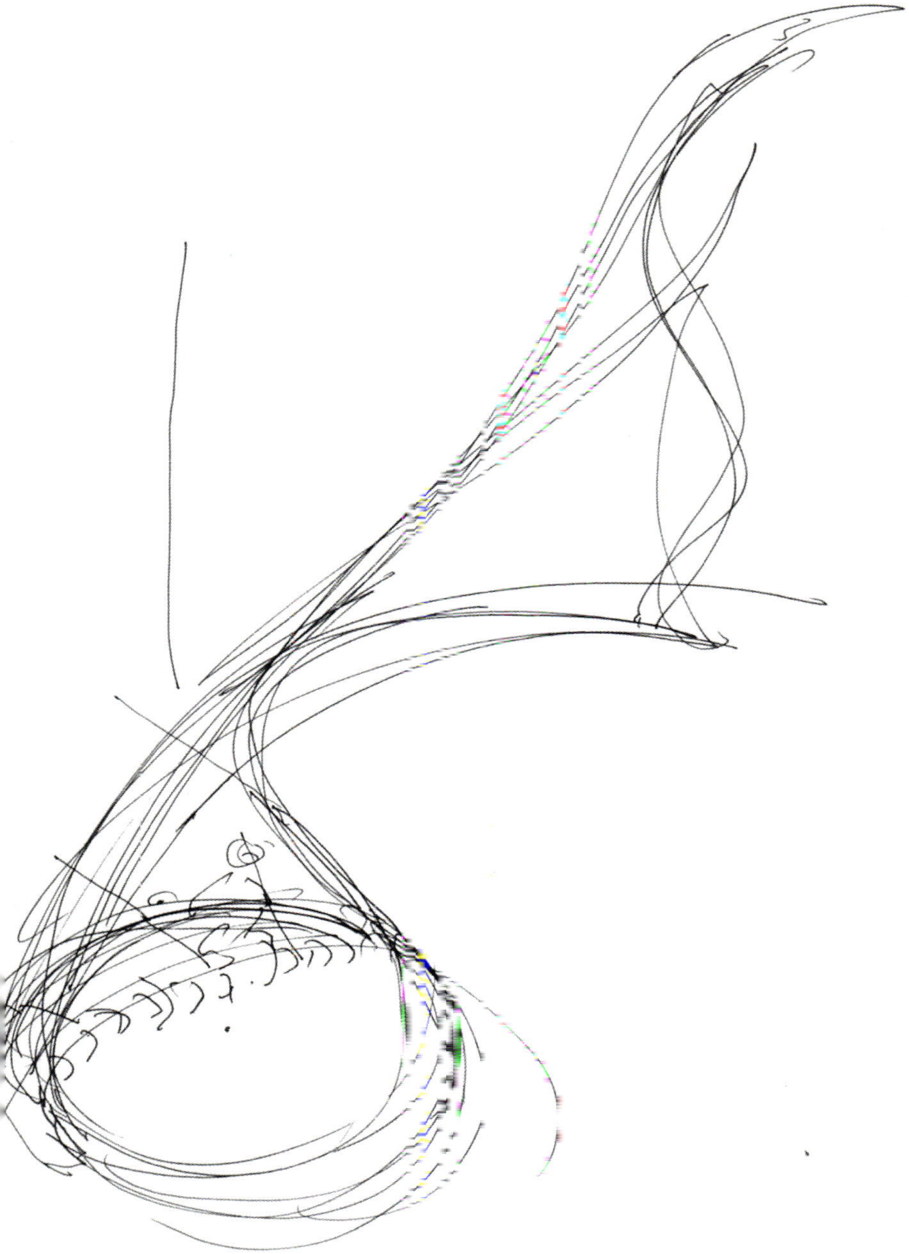

项目介绍

设计初期，设计师想要设计一个用风向标带动的旋转遮棚。与原来的形状相比，不同之处在于风。他们想使用一个风洞（wind tunnel），但因为成本的关系没有得到开发商的同意，因此，他们想了个办法。用一个风扇代替风洞，如果还有一根线和一支铅笔，也就能知道，这个办法是否可行。设计师用这个简单的办法设计出了许多形状。最后，他们设计了一个建筑原型。

设计细节

设计师们不能使用任何复合几何图形。因为他们必须把费用控制在合理范围之内。无论如何，他们还有些担心，是否有些装饰过分。但事实上很简单，设计师采用一个中央脊柱，即一根对称轴，根据它来设置垂直部分和削掉的部分。底座上面的架构有一层表皮，“鱼鳍”向下延伸，然后渐渐扩张，托起长椅。

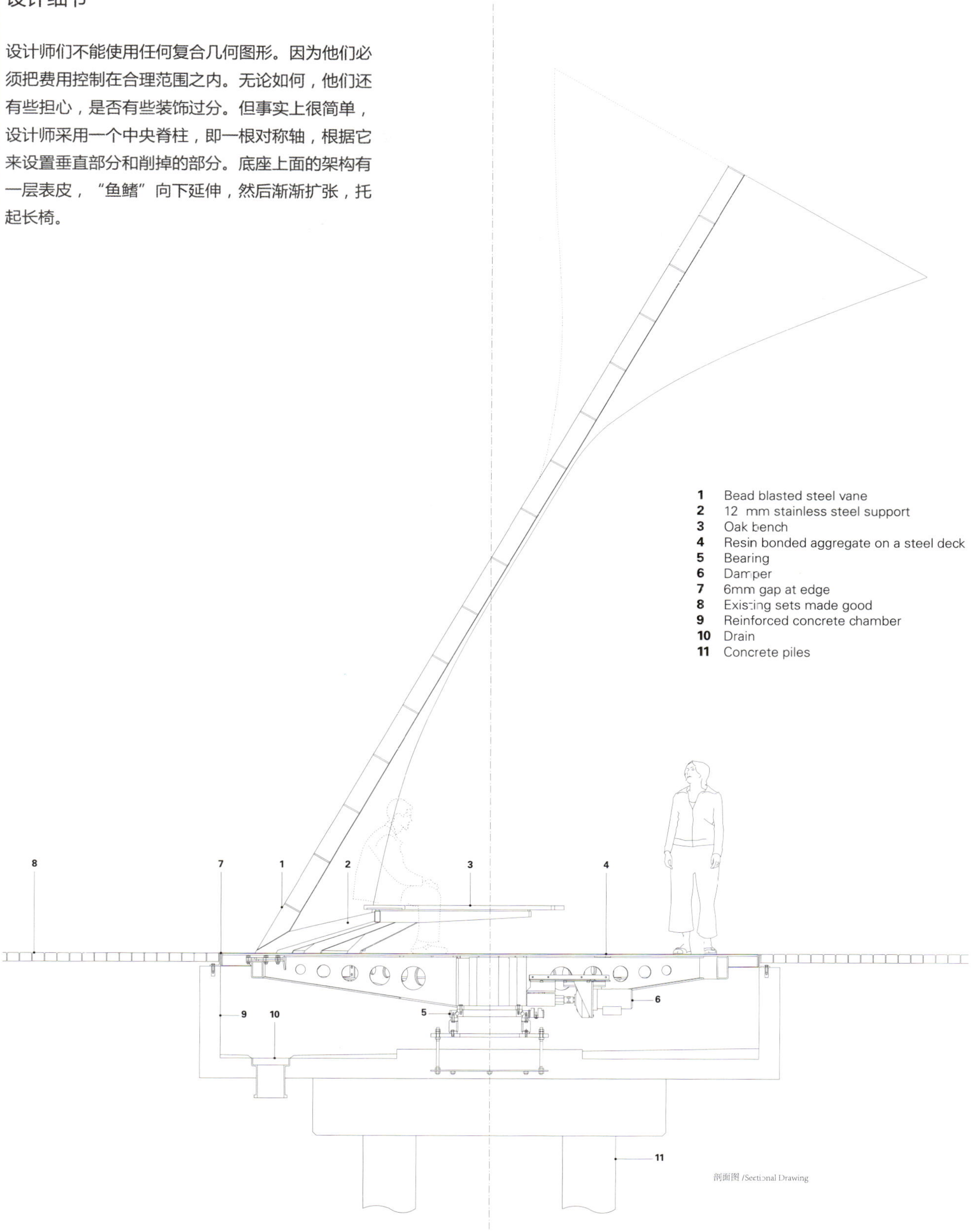

剖面图 /Sectional Drawing

Athletes Village

运动村

Project Location: Whistler, Canada
Architectural Design: Tom Barratt
Completion
Year: 2010
Client: Canada Olympic Committee

项目地点：加拿大
建筑设计：TOM BARRATT 设计事务所
竣工时间：2010 年
开 发 商：加拿大奥委会

Details Determines Sports Ground 细节成就赛场

Project Description

As an Olympic and Paralympic Village, universal accessibility was a driving objective of the design. All building entries, road crossings, site pathways, general site furnishings—even children's play areas—had to be universally accessible. Designers began to look more carefully at site furnishings as a way to both effectively achieve accessibility, and showcase art at the same time.

VILLAGE PARALYMPIQUE DE WHISTLER

BUS
STOP

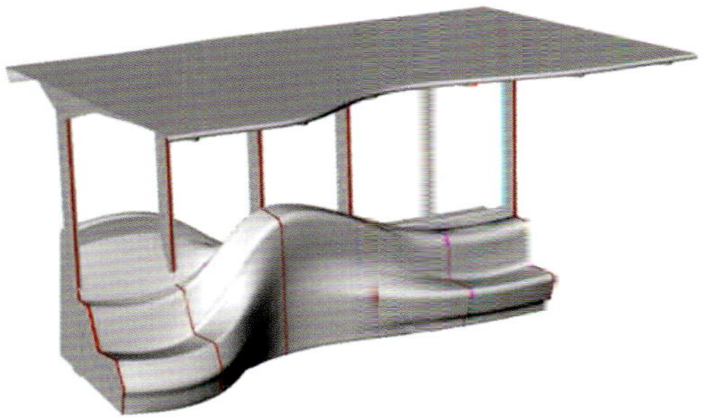

5 CORIAN shelters built with modular system of aluminum ribs lined with LED from structural rib through the CORIAN seating. Seating and rib profiles are re-used for each shelter, for ease in fabrication process, but are reconfigured to provide varied and unique shelters at each site.

Design Details

To bring a unique style to the Athletes' Village, designers purposely departed from the existing aging Whistler Village style and brought in some outside expertise. They collaborated with Urban Movement of New York (Architects, Body Movement Specialists) to produce the custom curvy corian benches, the bus shelters, and the multi-purpose bikes rack/stretching features. To produce the essential bear-proof garbage cans, designer retained an artist [illegible]senberg Design of Colorado. The cutting-edge LED lighting standards were supplied by the Schreder Group. Another concern was storm water management. Kerr Wood Leidal engineers designed a comprehensive storm water system.

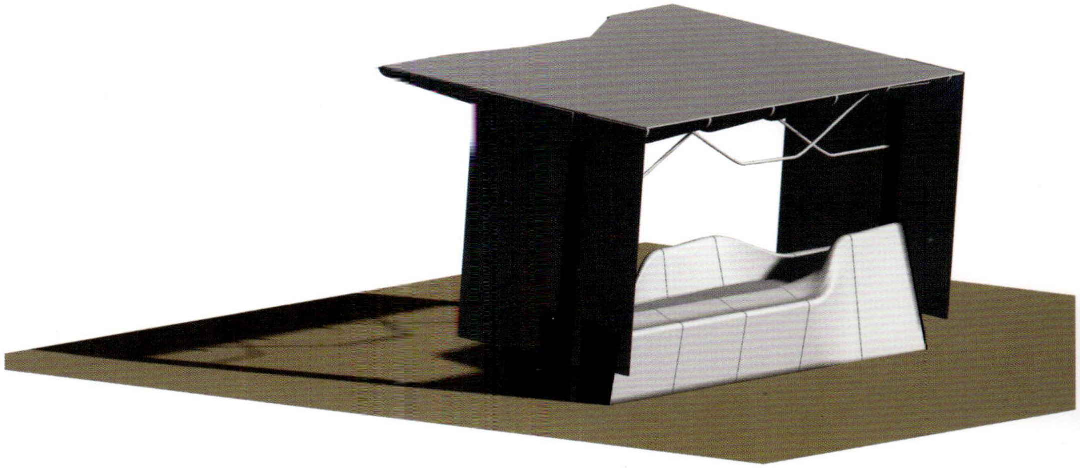

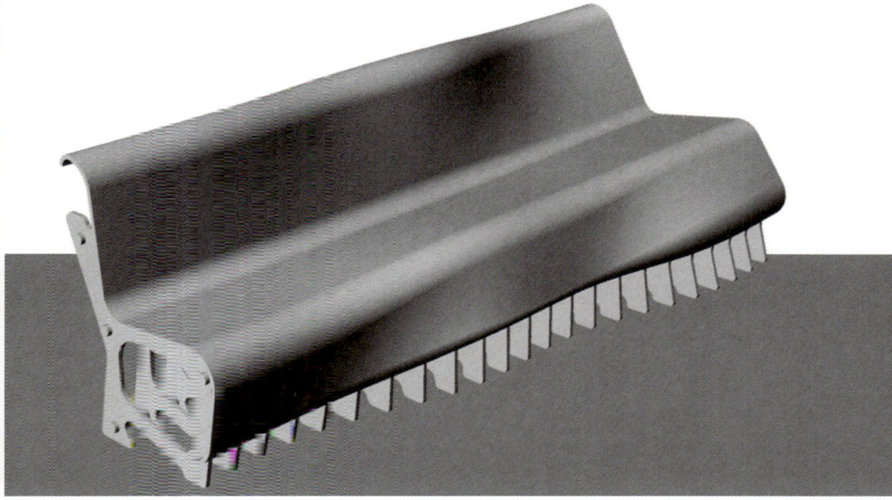
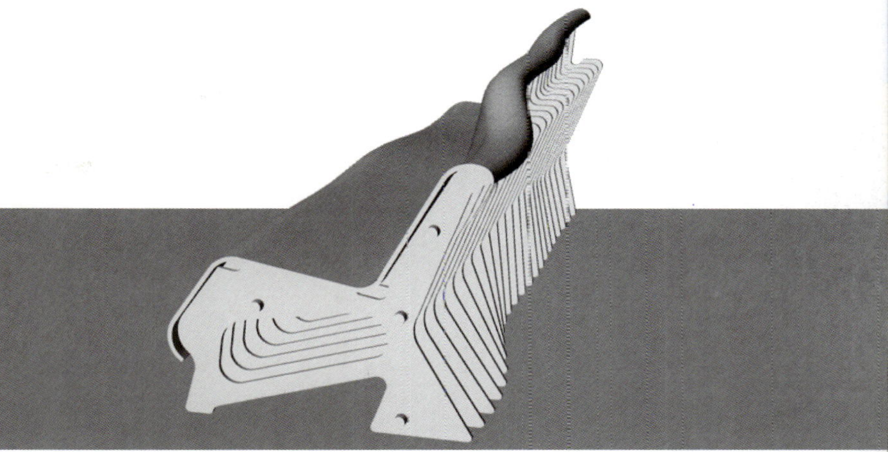
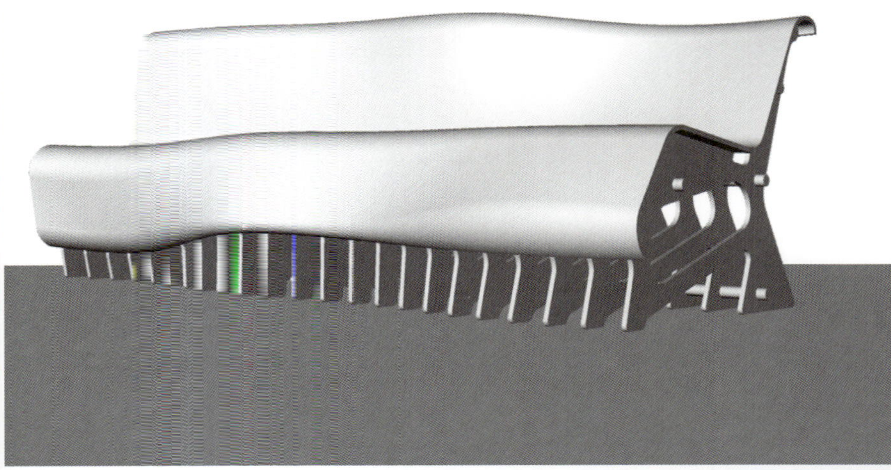
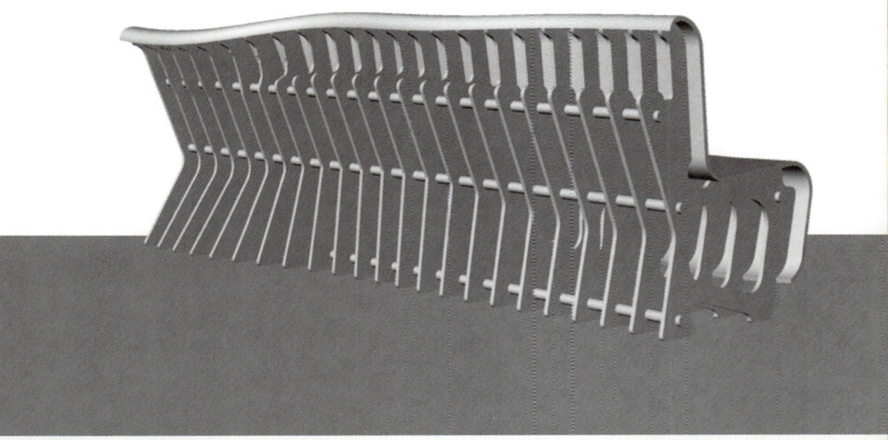

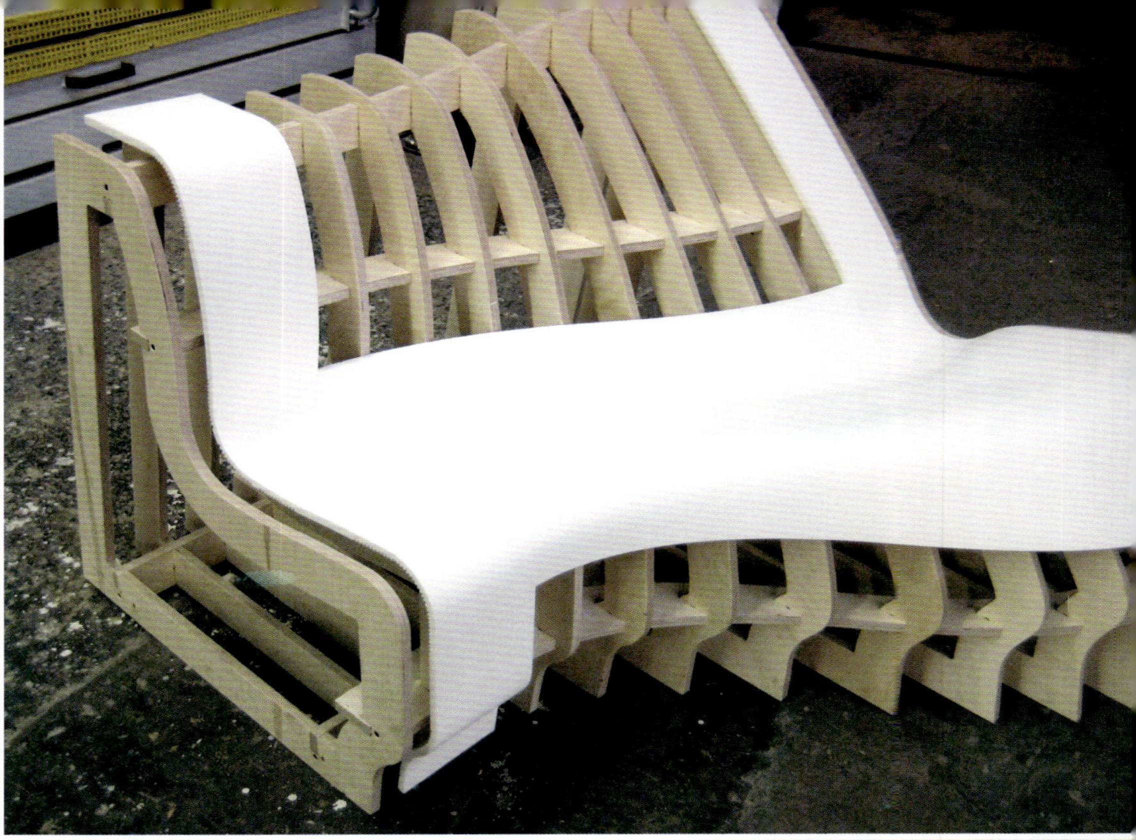

项目介绍

该地区作为奥运会和残奥会的奥运村，在设计理念上要突出“畅通无阻”的特点。因此所有建筑的入口处、道路的交叉口、行走的小径和奥运设施，甚至是小孩的游乐场所，都必须满足“畅通无阻”的原则。于是设计师们从奥运赛场上的设施着手，在满足畅通的同时，也与展示艺术理念结合起来。

设计细节

为了赋予奥运村一些独特的风格，设计师抛弃了惠斯勒村的一些陈旧风格，同时注入了一些新的元素。设计师与纽约建筑师合作，共同设计了预制可丽耐弯曲长凳、公共汽车候车亭，以及多功能自行车车架。为了制造一些必要的防爆垃圾桶，设计师还聘请了科罗拉多州的艾森伯格设计公司的一名艺术家进行设计。奥运村里顶尖的 LED 照明系统是由施雷德集团提供的。该项目另一个重要的设计理念就是雨水处理系统。Kerr Wood Leidal 的工程师们设计一个比较完善的雨水排除系统。

bike rack v1 - plan

bike rack v1 - side elevation

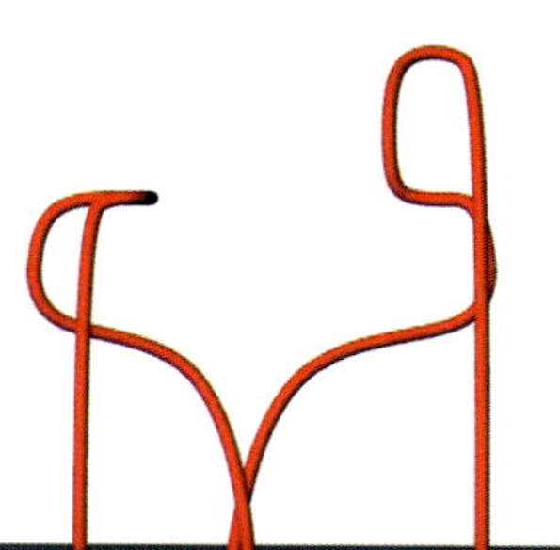

bike rack v1 - front elevation

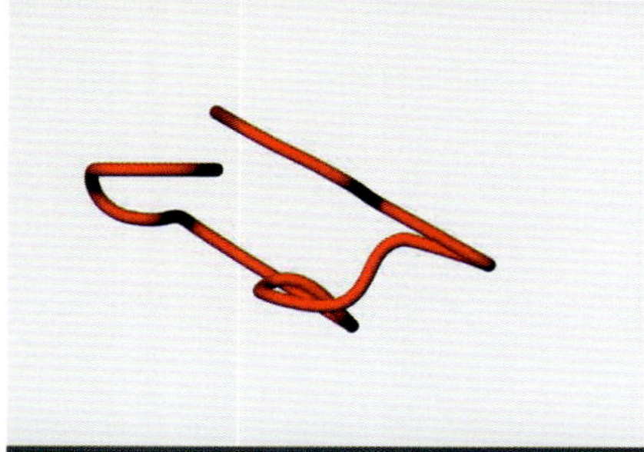

bike rack v3 - plan

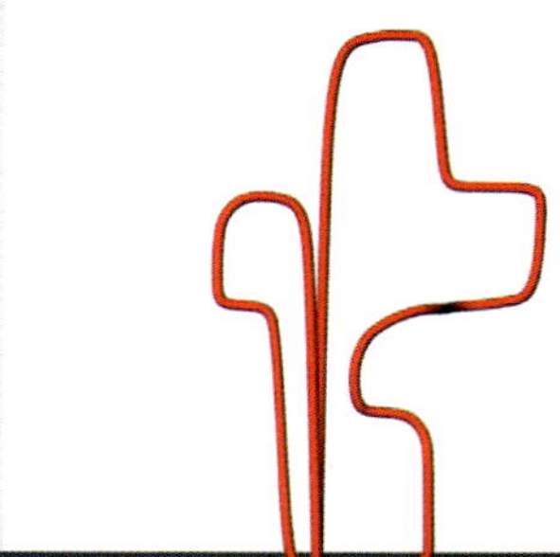

bike rack v3 - side elevation

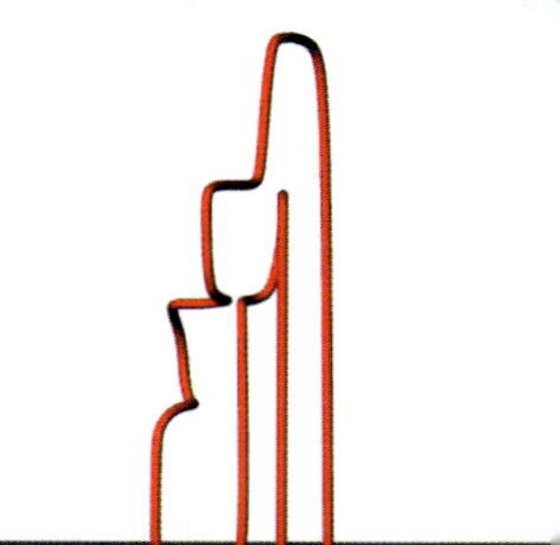

bike rack v3 - front elevation

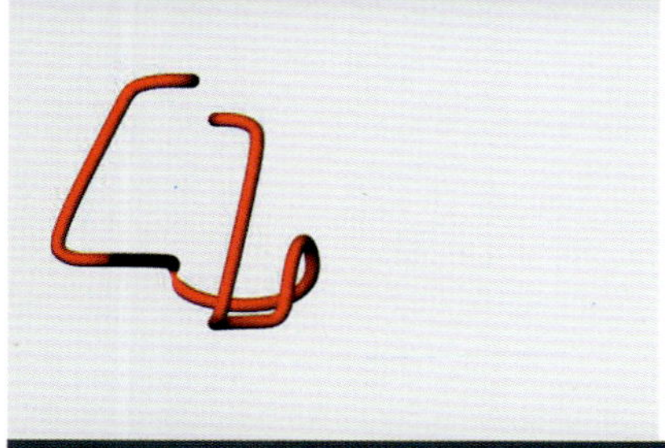

bike rack v2 - plan

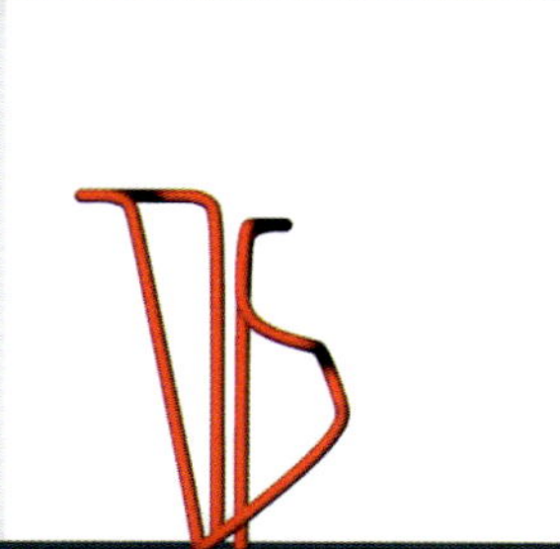

bike rack v2 - side elevation

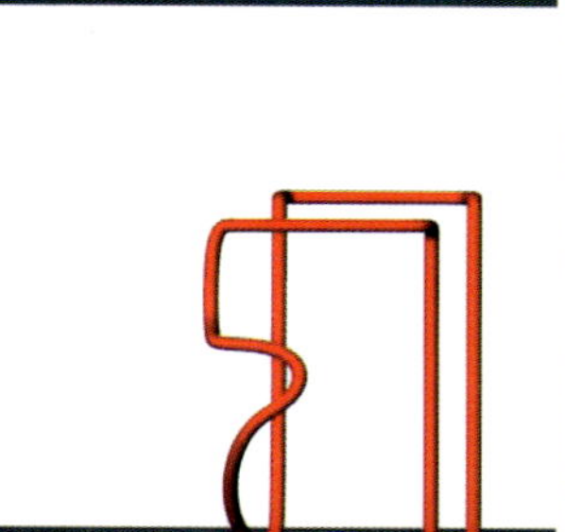

bike rack v3 - front elevation

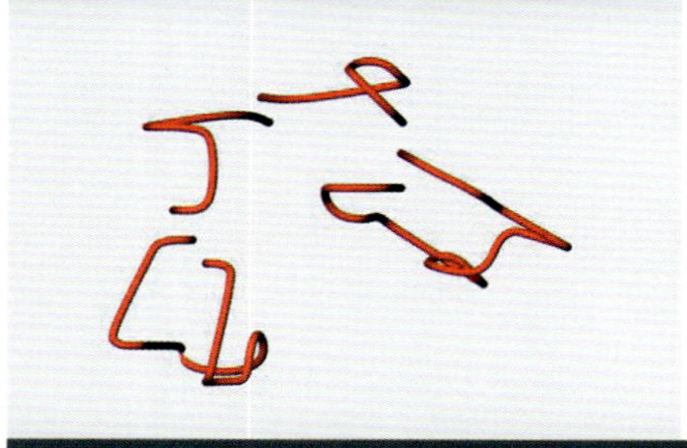

bike rack group - plan

bike rack group perspective 1

bike rack group perspective 2

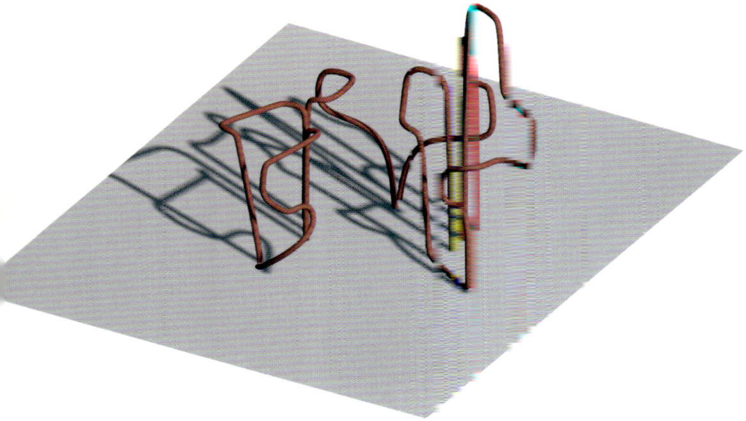

Boston Treepods

波士顿的人工树

Project Location: Boston Massachusetts, USA
Architectural Design: Influx Studio
Completion Year: 2011

项目地点：美国马萨诸塞州波士顿
建筑设计：Influx Studio 设计事务所
竣工时间：2011 年

An Inspiration From Bio-mimicry and Dracaena

生物仿生学与龙血树的灵感之作

Project Description

The aim of the project is to create, using biomimicry, an air cleaning and CO_2 catcher integrated urban device. Looking at nature we can learn from one of the most unique trees in the world, the Dragon Blood Tree (*Dracaena cinnabari*). Its branches at maturity produce an umbrella shaped crown optimizing its form to create a canopy that provides a maximum of shading surface. The way that its canopy allows the wind flow is showing us an intelligent form like design. The Treepod will be inspired by that, along with its branching structure in terms of storage and distribution of resources from ground to the canopy.

Design Details

The Treepod takes the Dragon tree like form to create an important canopy surface that will provide shadow, and that will host solar photovoltaic cells to harvest the energy necessary to power the air cleaning system and the urban lamp function. The canopy branching structure ends with a myriad of bulbs. They multiply the contact points between air and CO_2, serving as a filter. Working like as alveoli in a human lung, here is where the cleaning gaseous exchange takes place: an alkaline and environmentally friendly resin that reacts with air holding CO_2. When the CO_2 saturated resin reacts with water it releases CO_2 for storage, and then it could be used again in the same process.

NEW CHARDON ST
Gov't Center
Government Center Garage
CLEARANCE 13'-4"
NORTH 1A
CLEARANCE 13'-9"
93 SOUTH
Quincy
93 SOUTH
90 WEST
NEW CHARDON ST
NO TURN ON RED
NORTH 93
boston
TREEPODS
OK

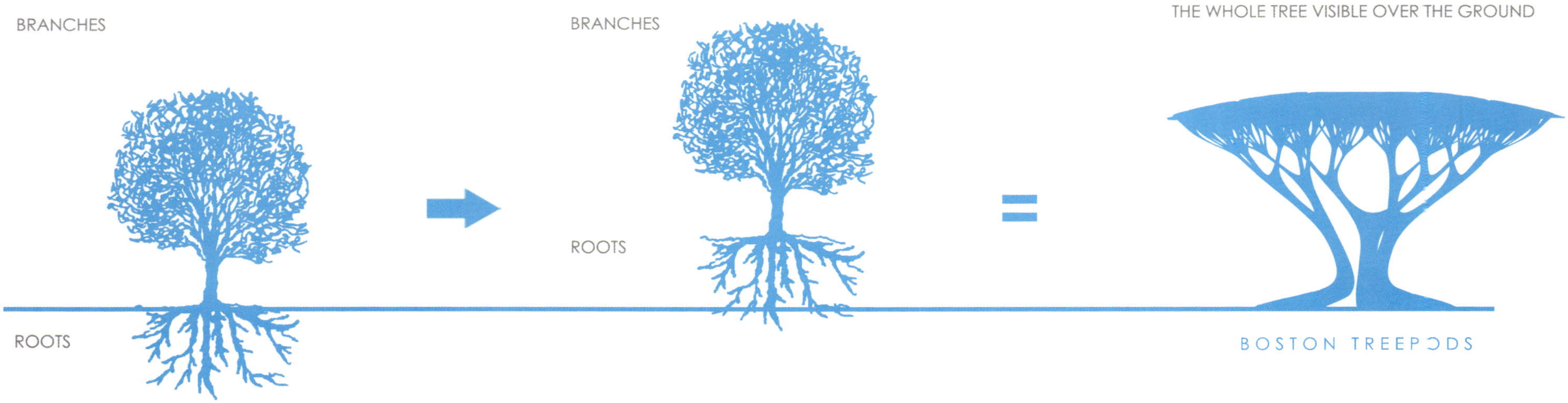
BRANCHES
ROOTS
BRANCHES
ROOTS
THE WHOLE TREE VISIBLE OVER THE GROUND
BOSTON TREEPODS

≥

ONE TREEPOD

100 TREES

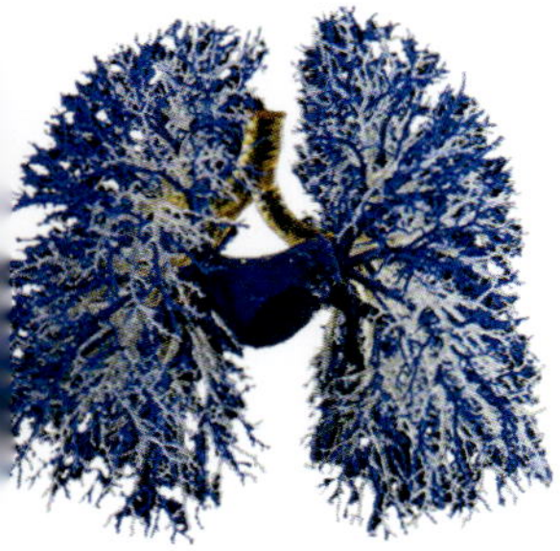

PULMONARY ALVEOLI
STRUCTURE

+

THE DRACAENA CINNABARI
OPTIMIZED CANOPY SHAPE

=

BOSTON TREEPODS

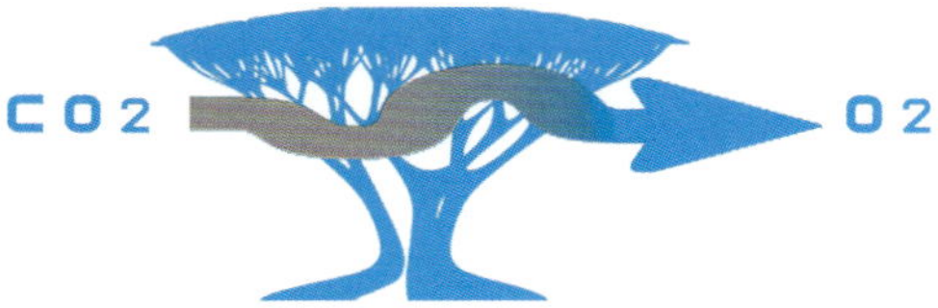

CO2 SCRUBBING ARTIFICIAL TREE

DRAGON BLOOD TREE LIKE SHAPE

LUNG LIKE ALVEOLI SYSTEM

SOLAR CANOPY / KINETIC ENERGY HARVEST

RECYCLED PET BOTTLES STRUCTURE

SUN SHADING / RAIN SHELTER

GLOWING LED LIGHTING

SOCIAL INTERACTION

TO BE INVOLVED

项目介绍

该项目利用了生物模仿科技，是一个可分离二氧化碳、净化空气的城建设施。其灵感来自于大自然中的龙血树（*Dracaena cinnabari*）。该植物成熟后的枝干会异常茂密，它们像大伞一样有遮挡的功能，同时也能加快风的流动。这种巧妙的设计，使其可以利用枝干结构向顶棚部分储存和传输从地面而来的资源。

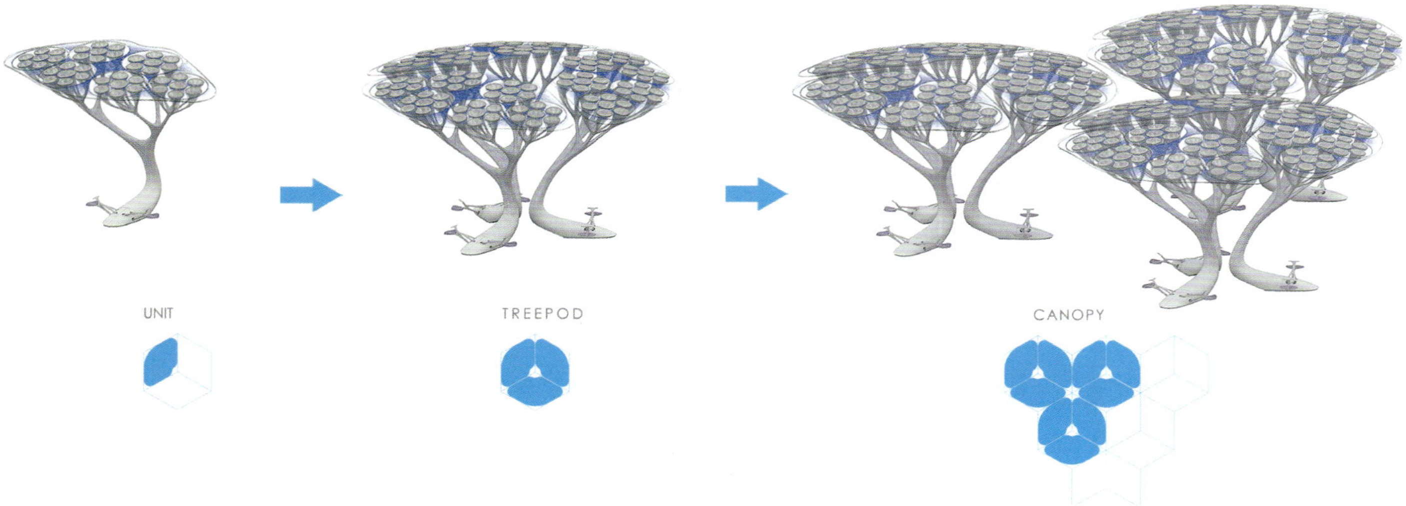

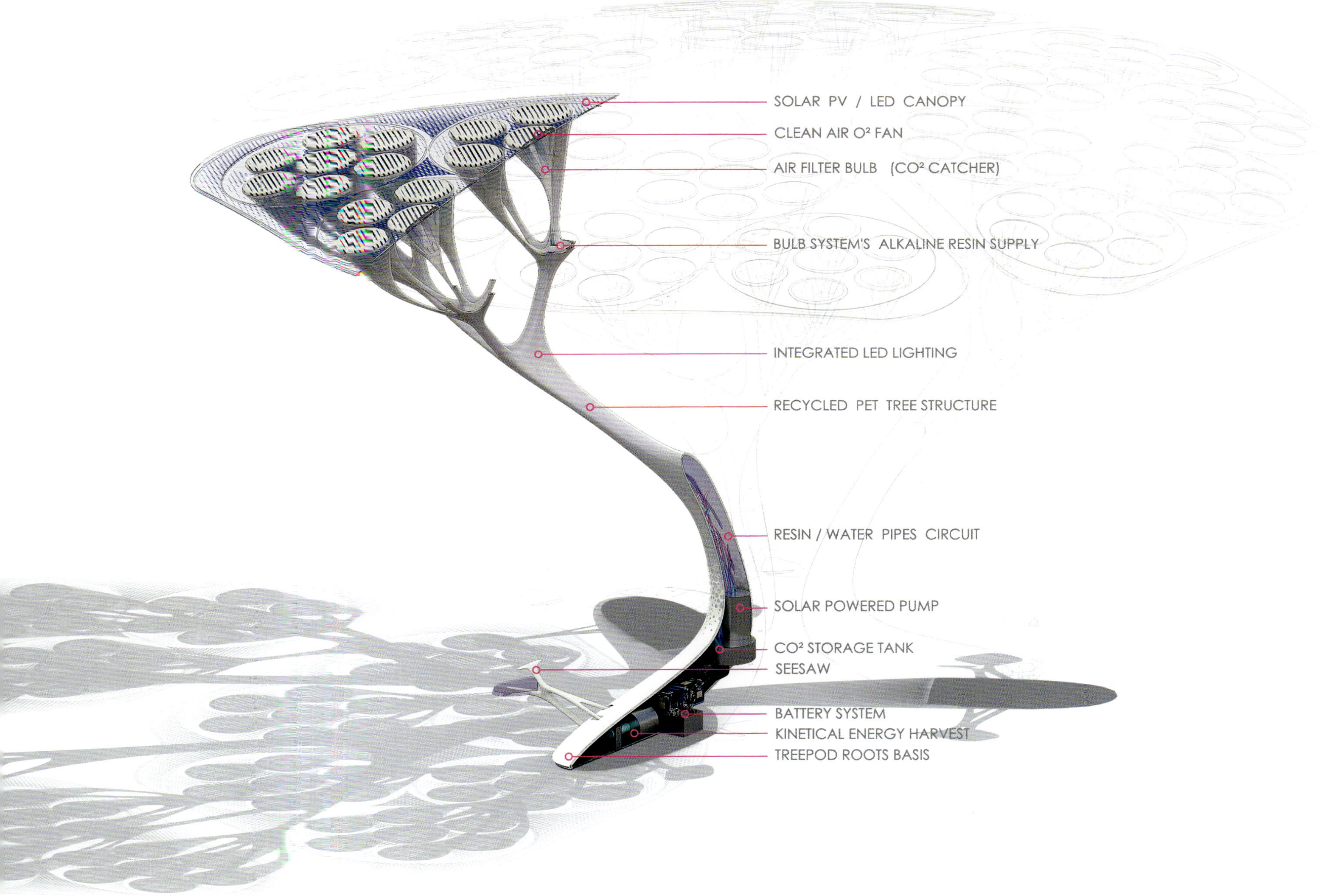
SOLAR PV / LED CANOPY
CLEAN AIR O² FAN
AIR FILTER BULB (CO² CATCHER)
BULB SYSTEM'S ALKALINE RESIN SUPPLY
INTEGRATED LED LIGHTING
RECYCLED PET TREE STRUCTURE
RESIN / WATER PIPES CIRCUIT
SOLAR POWERED PUMP
CO² STORAGE TANK
SEESAW
BATTERY SYSTEM
KINETICAL ENERGY HARVEST
TREEPOD ROOTS BASIS

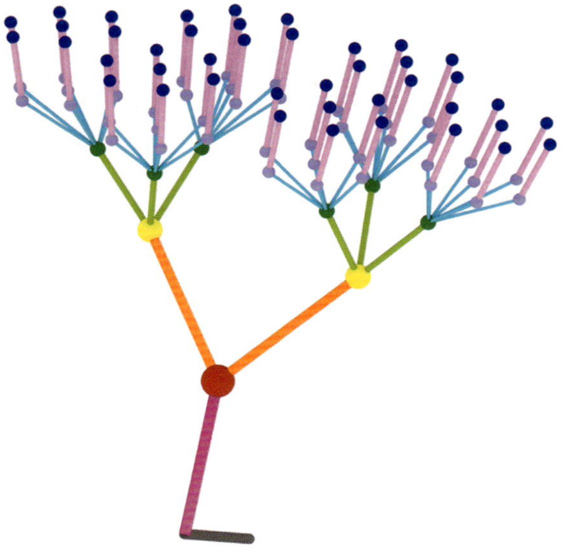

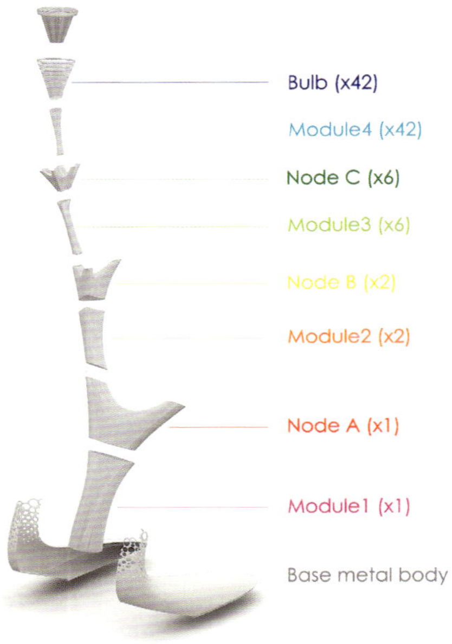

设计细节

人工树的“树冠”部分不仅可以遮阳，还利用了光电板发电，为过滤系统和城市照明灯提供能源。枝干结构的末端有无数个球茎，它们增加了与空气、二氧化碳的接触面，充当了过滤器的角色。如同人类肺部的气泡一样，环保型树脂与空气接触，将二氧化碳吸收。吸满二氧化碳的树脂与水反应后，二氧化碳将被释放出来，树脂又可以继续运作。

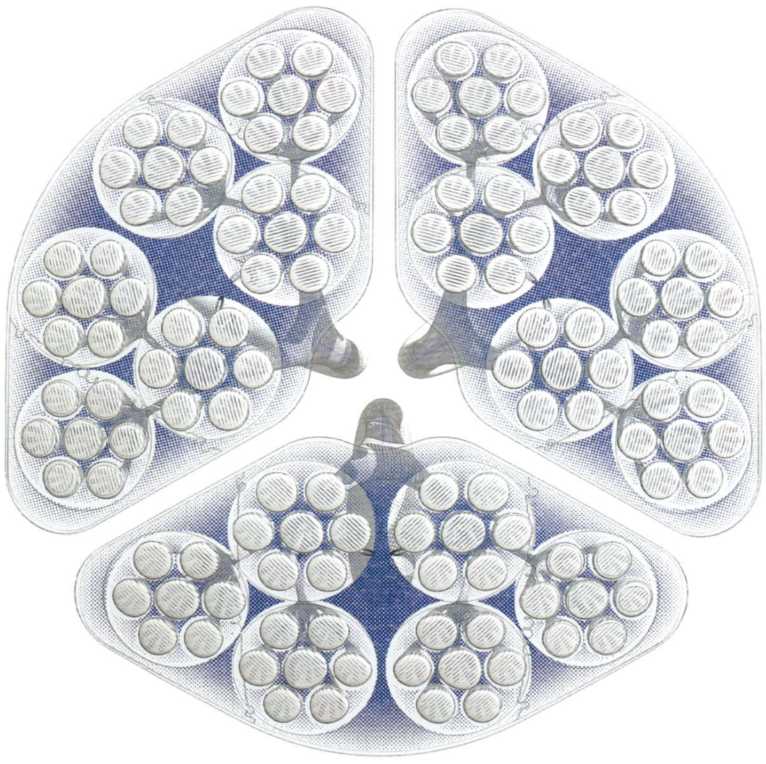

FAMOUS ITALIAN FOOD
NICO

Interior Decoration
室内装饰

La Familia

“家庭的温暖”

Project Location: Israel
Architectural Design: GORA art & Landscape
Completion Year: 2011

项目地点：以色列
建筑设计：GORA art & landscape 设计公司
竣工时间：2011 年

An Entrance of Infection Clinic in Malmo 马尔摩传染病诊所入口

Project Description

La Familia is what you first encounter as a visitor to the new infection clinic in Malmo. La Familia is a composition made up by 28 plastic sculptures, reinforced with fiberglass. Made up by various colors and shapes, La Familia – just like members of a family–is similar, yet different. Hence, they are perceived as they belong to each other, the whole and wide family with all its different needs and desires. When you arrive at the emergency department you are in a state of crisis. You put your trust into the professional skills of the medical team to give you the right treatment for your condition but you also have a need to experience emotional care, support from loved ones, the presence of sympathy and compassion.

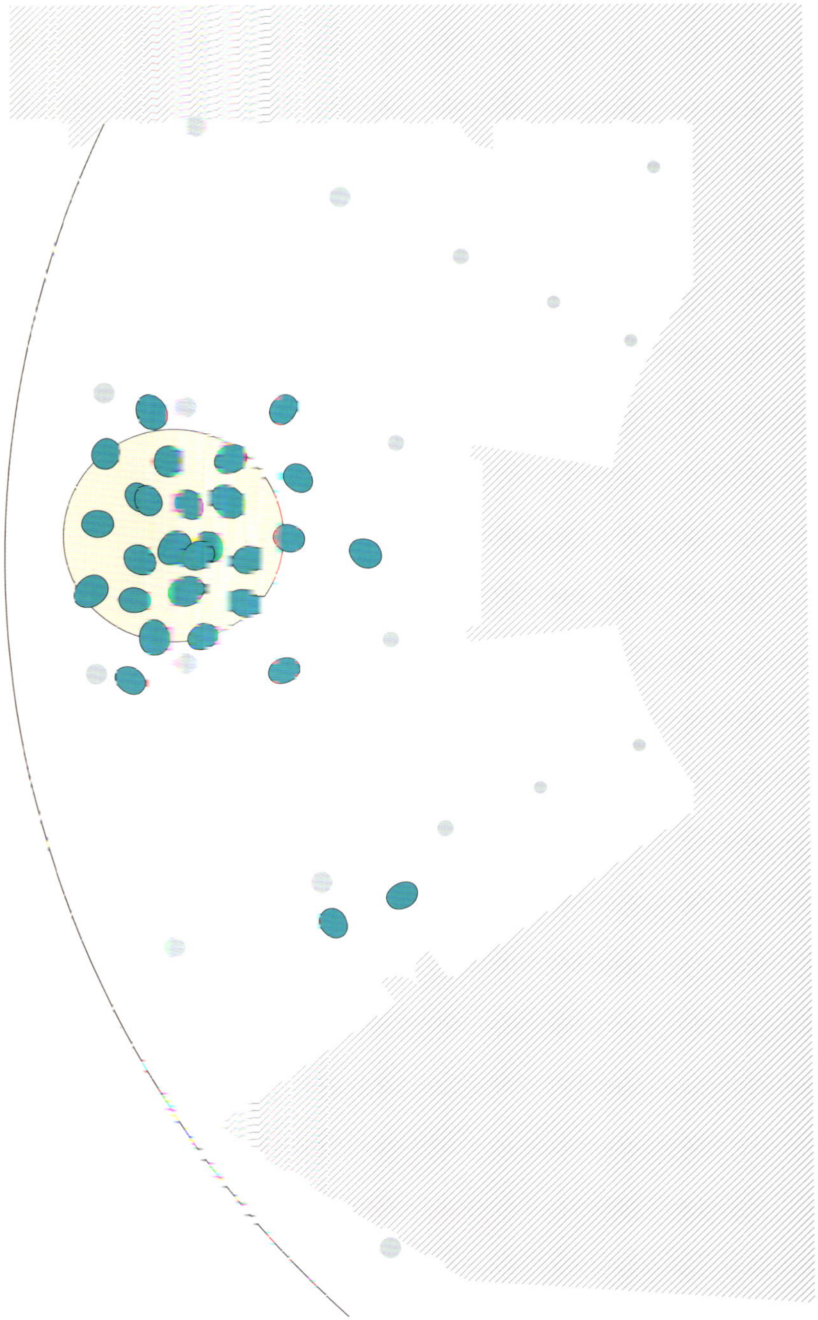

Design Details

La Familia embrace you even if you arrive by yourself, the sculptures are waiting for you at the entrance of the building. The colors and the shapes of the sculptures recall the connection to all the animals and plants living on this earth. The soft and irregular shapes are placed in the ceiling of the entrance hall. This family is large, warm and helpful. They meet you on your way out and on your way in. There is a large circular and elevated seating area under the sculptures where it is possible to sit down and spend some time with them. Some of the sculptures are scattered and some are gathered in groups. Together they form a unit that emits softness and tenderness. When it gets dark they emit light and bring along a feeling of safety and hope. Their form reminds of flowers, hearts or perhaps butterflies up in the ceiling and down on the ground. At the same time they emit comfort and playfulness.

项目介绍

La Familia 将会是你作为一个来访者，在马尔摩第一次遇见的传染病诊所。La Familia 的入口处有 28 个塑料雕塑作品，用玻璃纤维加以稳固。La Familia 运用各种颜色和形状，就像一个家庭里面的成员，虽有相似，却又不同。因此，他们察觉到自己也是属于家庭中的成员，自己要满足他们的不同需求和想法。若你在急诊科接受治疗，那么此时的你正面临生命危险。你将信任给了有专业技术的医疗队，以便他们可以帮你对症下药，但你也需要亲人们从情感上给予的关怀和支持，并从表情上给予的同情和怜悯。

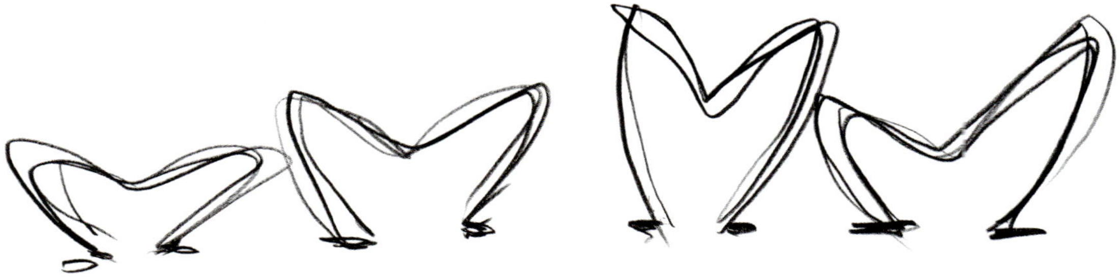

设计细节

即使是你独自来到这里，La Familia 也会给你热情的拥抱。门口的雕塑正等待着你。雕塑的颜色和形状好像是自然界里生长的动物和植物。柔软和形状不规则的雕塑设计在门厅的天花板上。这个大家庭是热情且乐于助人的。它们在你进出门的时候恭候着你。雕塑下方有一个大的圆形和高台休息区，方便你随时坐下以便打发一些时间。有些雕塑是分散的，有些则成群聚集在一起，它们统一起来，给人一种柔软亲切的感觉。每当夜幕降临，这些雕塑便开始发光，给人一种安全感，并带来希望。他们的形状让人想起了花和心，或者是蝴蝶飞舞于天花板和地板之间。与此同时，它们也具有舒适性和趣味性。

Zara Installation

Zara 装置

Project Location: Milan, Italy
Architectural Design: UNStudio
Project Area: 50 sq.m.
Completion Year: 2010
Client: Zara Shop
Building materials: Coated Aluminum Panels

项目地点：意大利米兰
建筑设计：UNStudio 建筑事务所
项目面积：50 平方米
竣工时间：2010 年
开 发 商：Zara 品牌店
建筑材料：涂层铝板

A Hollow Cone Structure 空心锥建筑

Project Description

The temporary installation has been designed by UNStudio for Zara on the occasion of the Salone Internazionale del Mobile. The installation consists of a hollow and accessible cone made from coated aluminum. Functionally, it is a cross between a sculpture, an item of furniture, and a pavilion. Its position is directly behind the entrance of the shop, at a prominent spot in the atrium where many period details of the shop's previous incarnation as a cinema are preserved. The installation responds to this situation in both shape and materialization.

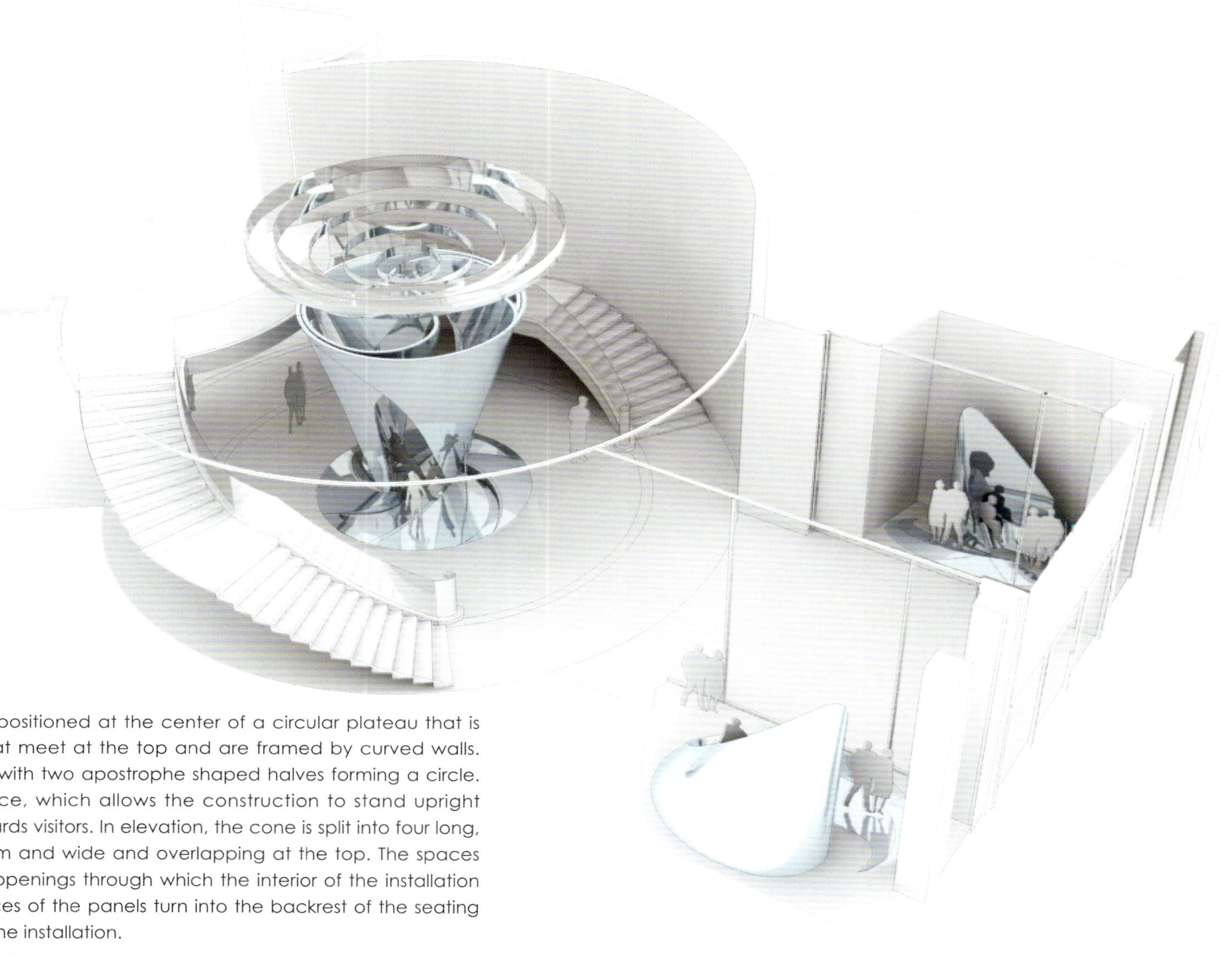

Design Details

Both installation and chandelier are carefully positioned at the center of a circular plateau that is semi-enclosed by two sweeping staircases that meet at the top and are framed by curved walls. In plan, the installation is centrifugal in shape, with two apostrophe shaped halves forming a circle. Sweeping lines form the pedestal of the piece, which allows the construction to stand upright independently, present an inviting gesture towards visitors. In elevation, the cone is split into four long, convex panels which are narrow at the bottom and wide and overlapping at the top. The spaces between these glossy, curved panels are the openings through which the interior of the installation is accessed. On the inside, the concave surfaces of the panels turn into the backrest of the seating elements that are integrated at the bottom of the installation.

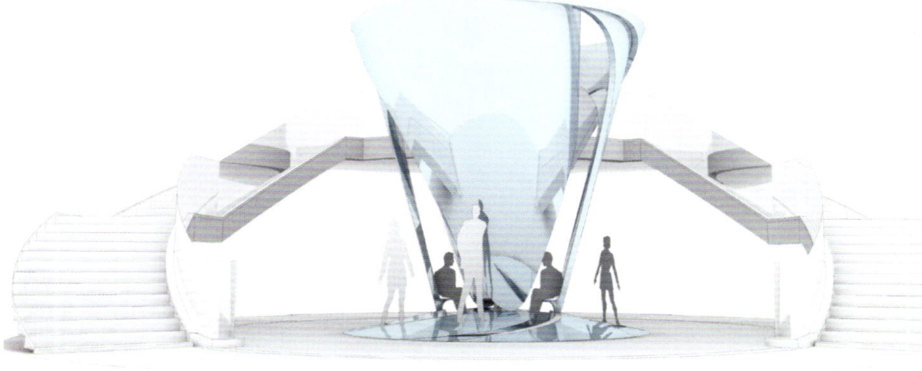

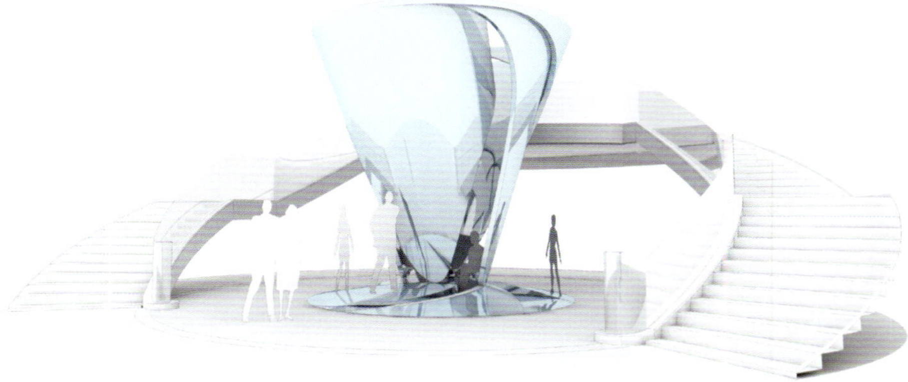

项目介绍

Zara 装置是在米兰国际家具展之际，由 UNStudio 建筑事务所设计的。该装置是用涂层铝板制成的一个空心锥。从功能上看，它可以是一个雕塑、一件家具、一个展馆或装置。它位于 Zara 品牌店的入口处，也就是店里最显眼的中庭，而中庭里还保留了许多有关店铺的前身为一家电影院的细节部分，它的设计在形状和材料方面都反映了这一背景。在里面，凹面的面板形成的座位与该装置的底部融合在一起。该项目的面板将顾客与忙碌的环境分离，同时把他们引入 Zara 品牌店中。

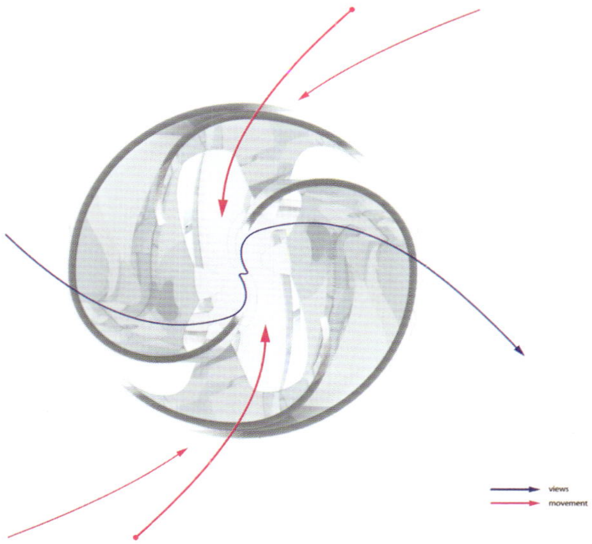

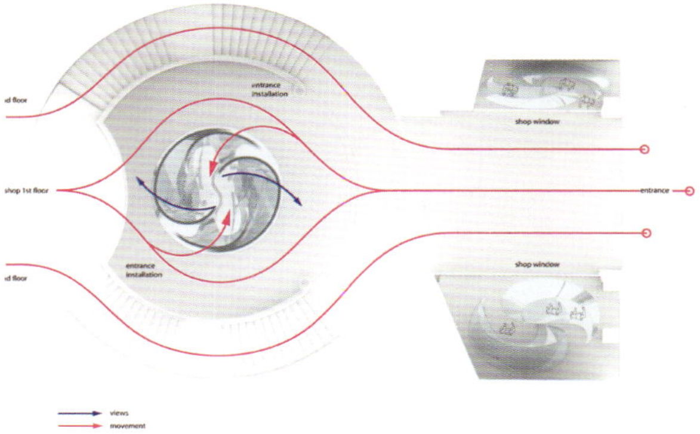

设计细节

Zara 品牌店的装饰和吊灯都被精心设计在圆形突出部分的中心，并用两边曲折的楼梯围绕在它周围，使它成为半封闭式，顶部便是弯曲的墙。在设计中，该项目为离心形状，并用两个一半的撇号形成一个圆形。基架弯曲的线条让这一装置独自矗立在展厅之中，向顾客展示着欢迎的姿态。从该建筑的正面图来看，这个圆锥形建筑被分成四个长且凸圆的面板，底部窄小，顶部渐宽且重叠。充满光泽且弯曲的面板之间的空隙便是这个室内装置的开口处。室内凹面面板变成了与底部融为一体的座位区靠背。

Armani Staircase NY

纽约阿玛尼品牌店内楼梯

Project Location: New York, USA
Architectural Design: Fuksas Design
Project Area: 2,800 sq.m.
Completion Year: 2009
Client: Giorgio Armani Group
Photography: Ramon Prat
Building materials: Steel, Plastic Layer, Wood Panel

项目地点：美国纽约
建筑设计：Fuksas 设计公司
项目面积：2,800 平方米
竣工时间：2009 年
开 发 商：阿玛尼集团
摄　　影：Ramon Prat
建筑材料：钢板、塑料层、木板

Radical Staircase Design 辐射状楼梯设计

Project Description

After Hong Kong Chater House and Tokyo Ginza Tower, Fifth Avenue completes the trilogy of the Armani Stores designed by Doriana and Massimiliano Fuksas for the famous designer. Situated in the centre of New York, in one of the world's well known streets, the project takes up the first three floors of the two buildings located between 5th Avenue and 56th Street. Besides the basement, the showroom develops on four different levels and it is conceived as a single space, without clear distinctions, a space in harmony connected with the power generated by the vortex that is the staircase. The heart of the building is, in fact, epitomized by the staircase. Structure in rolled calendar steel (made in Italy) and clad in a plastic layer highlights its exceptional sculptural presence. It is an entity that is almost impossible to convey in terms of any normal geometric shape that originates from a vortex with great dynamism, surrounded by the different levels that accommodates the Armani world.

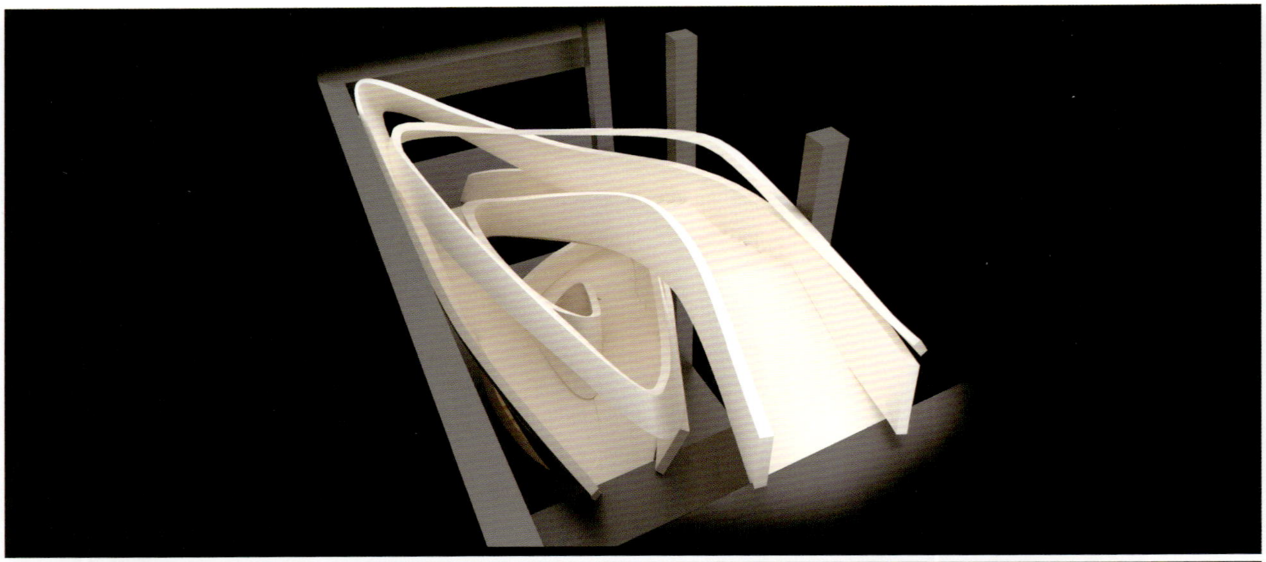

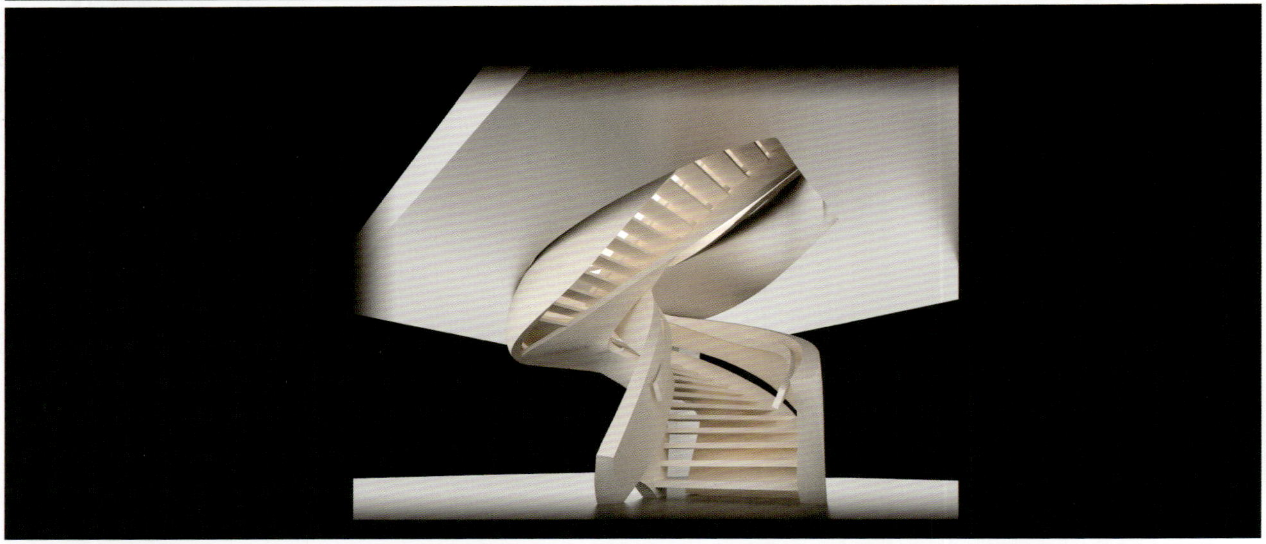

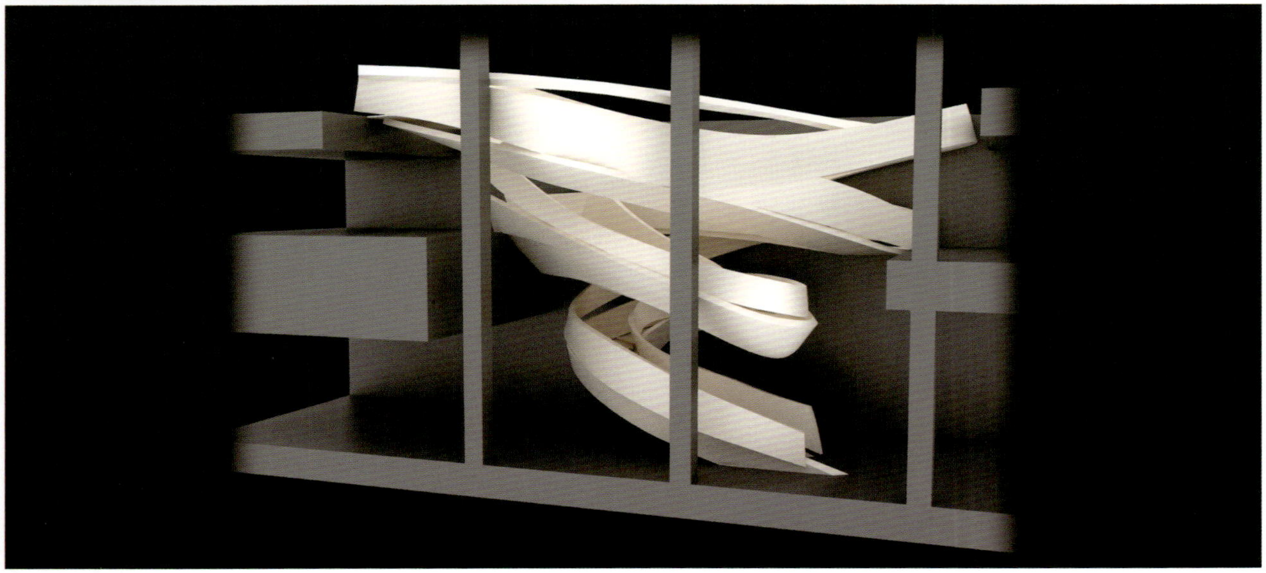

Design Details

The movement of the ribbons that constitutes the staircase, skimming each floor, disenchants the possibility to recognize the geometry. The general layout of every floor develops according to the different flexures of the ribbons, creating a space controlled by the vortex. No element is extraneous to the internal dynamism, not even the external facade; even if it is lined up to the rigid orthogonal stitch of Manhattan, simulating the movement through images and shades, projected on a set of LED threads. This screen, other than being the projection to the outside of the internal space, it is also a particular tribute to New York City, the inescapable necessity to compare its modernity and its dynamism. The fluidity of the internal space is rendered by the wall of continuous threads, which are realized with lacquered wood panels.

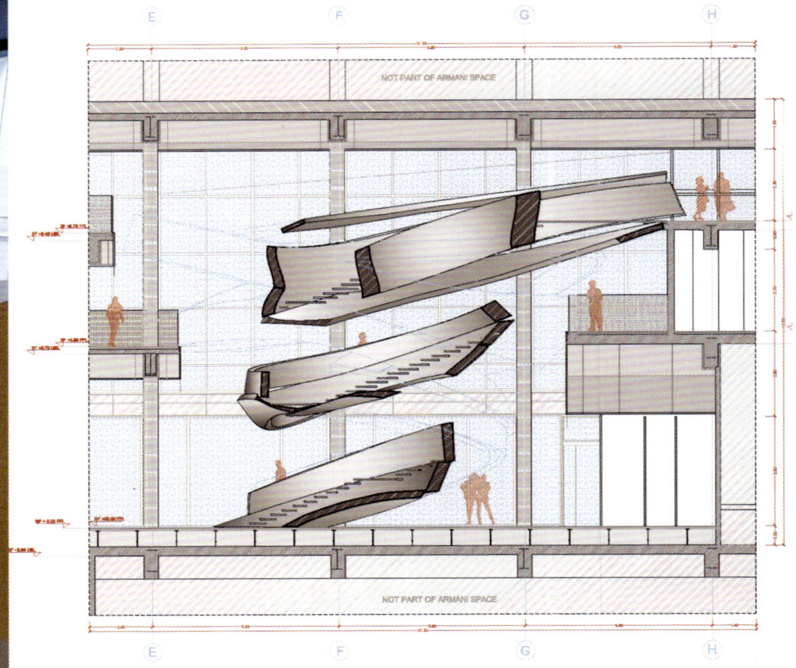
E
F
G
H
NOT PART OF ARMANI SPACE
NOT PART OF ARMANI SPACE
E
F
G
H

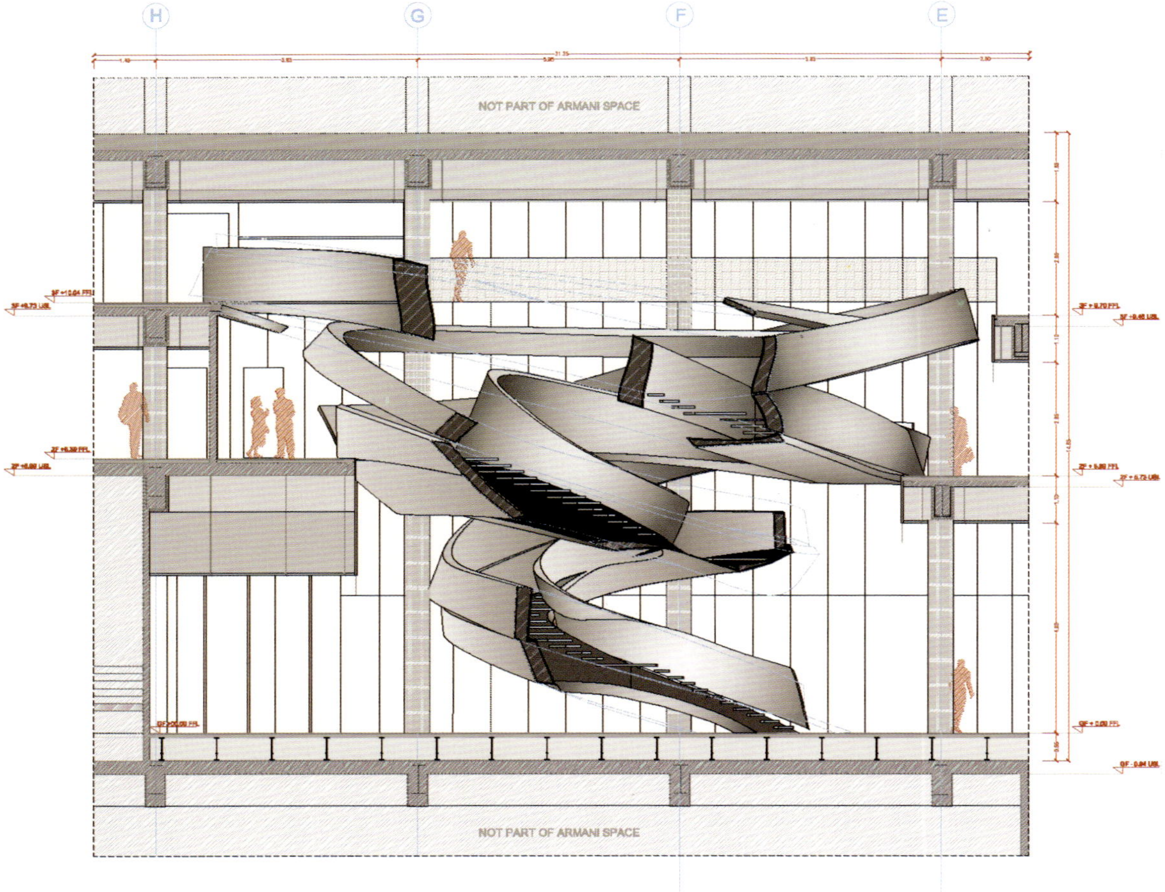
H
G
F
E
NOT PART OF ARMANI SPACE
NOT PART OF ARMANI SPACE

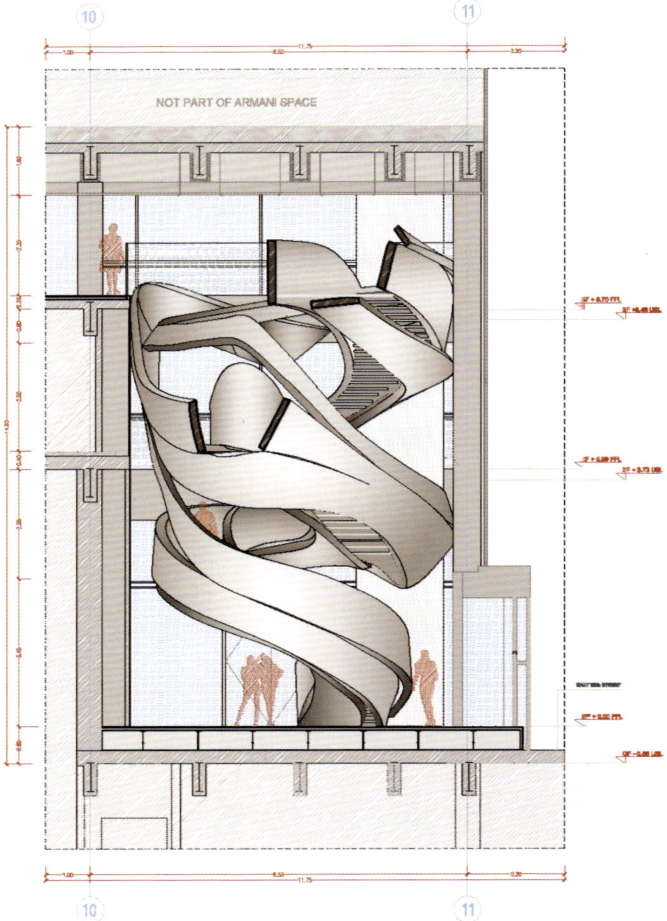
10
11
NOT PART OF ARMANI SPACE
10
11

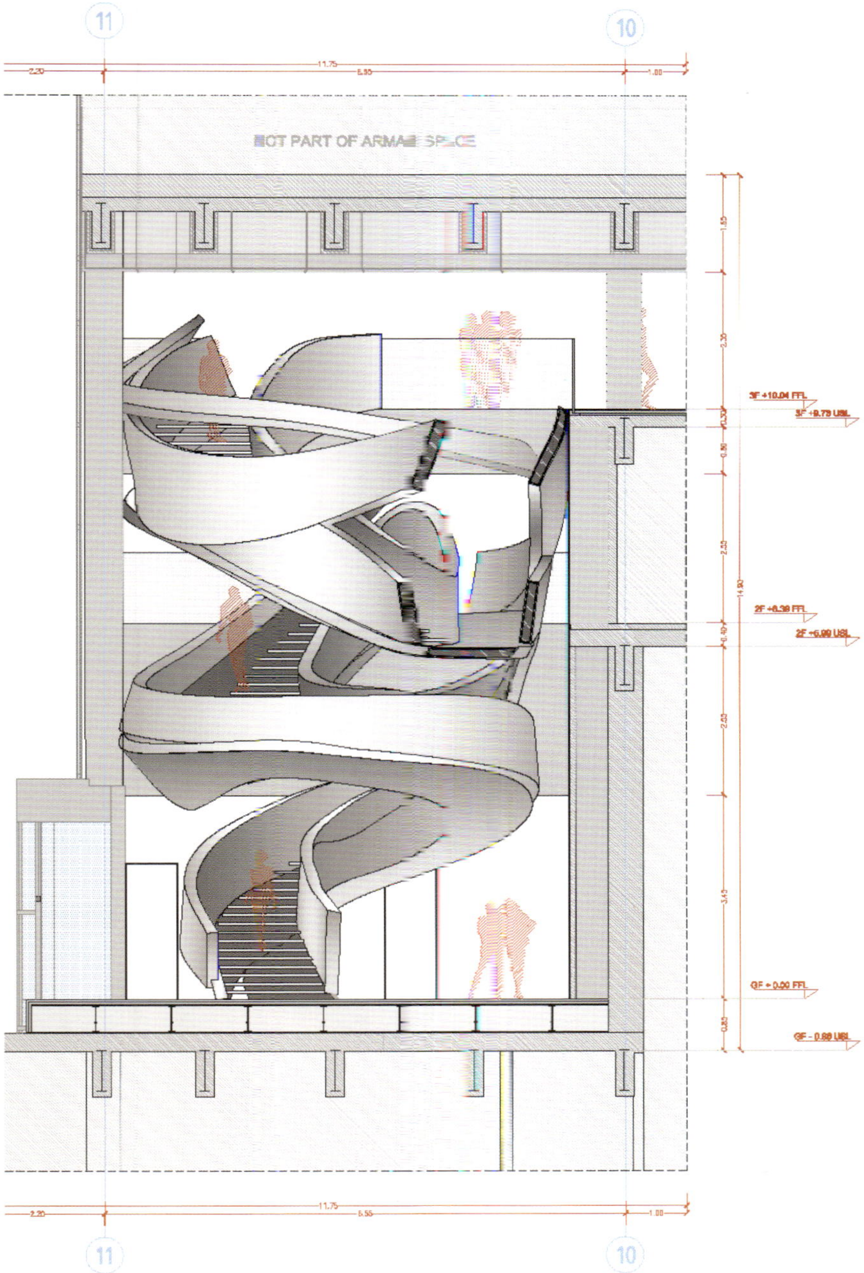

项目介绍

继香港和东京后，国际知名品牌阿玛尼又在纽约第 5 大道开店，完成了“设店三部曲”。阿玛尼品牌店是由意大利设计公司 Fuksas Design 设计的。阿玛尼品牌店位于第 5 大道和第 56 大街交汇处的两座建筑的顶三层。除了地下室，展示厅位于 4 个不同的楼层上，其核心点是一座连接各层的楼体。楼体采用辐射状的钢格板设计而成，表面用塑料雕塑。该项目是一个弯曲、独立的元素，既可以构成复杂的形状，也可以形成简单的几何轮廓，就像充满了极大动感的旋风一般，周围则全部是阿玛尼的世界。

设计细节

在楼体的带状运动中，每一层都构成了流动的形态，从两个不同角度看其形态也不同。内部空间的流动性效果是以单色漆粉刷的木板来完成的。内部设计的每一个要素，从展览室到寄存室，从桌子到扶手椅，都跟随着楼体的运动概念而变化，成为旋风中的一部分。在琥珀色“面纱”下的餐厅里可以一睹第 5 大道和中央公园的景色。色彩和材料的使用和展厅的其他空间保持一致，但意味却不同。地板上的一行灯光作为视觉上的幕布照射到餐厅的入口，让过道变得生机盎然，很像是戏剧空间。

Neil Barrett Flagship Store

尼奥・贝奈特旗舰店

Project Location: Tokyo, Japan
Architectural Design: Zaha Hadid Architects
Completion Year: 2008
Client: Neil Barrett
Building materials: Corian Finish

项目地点：日本东京
建筑设计：Zaha Hadid Architects 设计公司
竣工时间：2008 年
开 发 商：尼奥・贝奈特
建筑材料：可丽耐材料

The Dualism Between Male and Female 男女双重性的结合理念

Project Description

Rather than defining a single room or space, designers create a circular passage allowing the customer to experience the space in multiple ways and interpretations. Furniture staged in key points throughout the store creates the spatial concept of a narrow enclosure changing to an open condition. In two formal elements the design shifts between architecture and sculpture, where a compact mass of surface layers unravel and fold to form the shelving display and seating. The emerging folds will be used as display area for the NB accessory collection.

Design Details

The concept of the store plays with the complementary characteristics and the related dualism between male and female. This is echoed in the furniture design on both floors through the formal language and tactile quality of the materials used. The furniture piece on the ground floor is designed as a strong, masculine and dynamic form whilst the piece on the first floor enunciates femininity through more fluid contour lines. The first floor is designed in a more playful manner creating different zones that maintain the perspective view between them. This interplay between male and female is followed through in the general aesthetic concept, setting the furniture piece with a white smooth Corian finish against the raw fair faced concrete surface of the rest of the space. This is further accentuated through the contrast in color and finish of the white matt furniture finish against a black glossy floor.

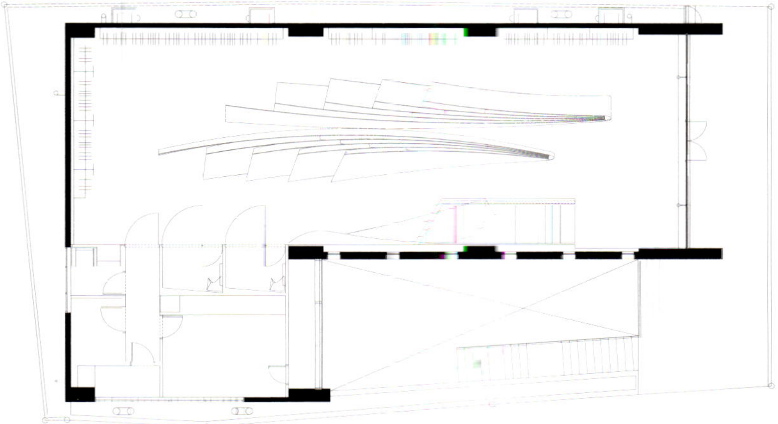

平面图 /Site plan

项目介绍

设计师创造的循环通道，让顾客以多种方式来体验这个空间，而不是定义一个单人房间或某个空间。整个卖场上，家具的摆放成为关键点，从一个狭窄的设计转变为一个开放式空间。设计师采取两种正式元素，使设计在建筑与雕塑间自由转换，表层紧凑的块状结构散开和折叠，形成倾斜的显示器和座位区。折叠的部分将作为 NB 件的展示区。

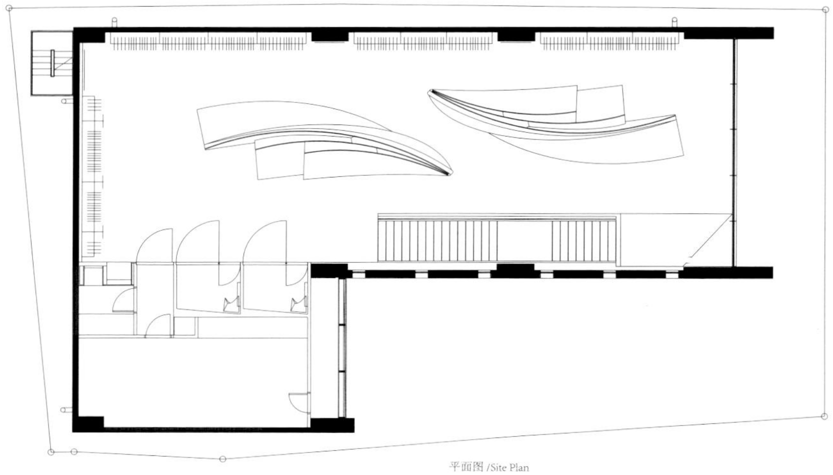

平面图 /Site Plan

设计细节

该项目的设计理念与现代特征相结合，与男女双重性理念相结合。这一理念在家具的设计方面，通过正式的语言和能触知的材料，在上下层再一次重复。一楼的家具设计是强健的男性化动态形式，然而二楼的设计通过更流畅的轮廓线条，清楚地表达了女性气质。一楼以一种更加俏皮的设计分隔出许多不同的空间，从任意一间都能透视到其它空间。通过采用一般的审美观，完成男性美与女性美的过渡。家具的设计采用白色可丽耐材料，与其它混凝土表面的空间设计进行对比。白色亚光家具与黑色光泽的地板进一步加深了色彩的对比度。

Gravity's Loom

地心引力织机

Project Location: USA
Architectural Design: Ball-Nogues Studio
Completion Year: 2010
Photography: Benjamin Ball
Building materials: String, Dye

项目地点：美国
建筑设计：Ball-Nogues Studio 建筑事务所
竣工时间：2010 年
摄　　影：Benjamin Ball
建筑材料：字符串，染料

Vibrantly Colored Hanging Strings 彩色字符挂串垂饰

Project Description

Indianapolis, IN, The Indianapolis Museum of Art announced that Los Angeles-based Ball-Nogues Studio will create a site-specific, architectural installation as part of the IMA's Efroymson Family Entrance Pavilion series. Ball-Nogues Studio's installation will be on view in the IMA's main entrance from September 3, 2010 to March 6, 2011.

Design Details

Gravity's Loom was composed of an array of vibrantly colored hanging strings that spanned the entire museum lobby and generate the appearance of a softly spiraling gossamer surface. This surface twisted, contorted, and spiraled downward through the atrium, transforming the architectural space and re-choreographing the flow of visitors to encourage new interactions with the museum. The strings of Gravity's Loom were painted to represent the imagined plan for a traditional Baroque ceiling pattern—a three dimensional volume that blurred into billows of color and then snapped into a focused geometry, depending on the viewer's vantage point.

项目介绍

由洛杉矶 Ball-Nogues Studio 建筑事务所设计的美国印第安纳波利斯艺术博物馆的艺术装置，它位于在国际矿物学会的埃弗罗姆森家族的入口处这一艺术装饰将于 2010 年 9 月 3 日 ~2011 年 3 月 6 日间完成。

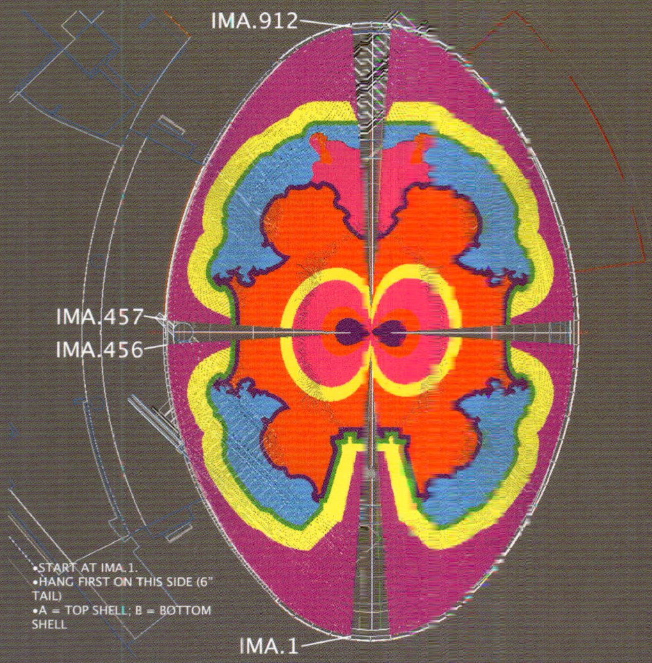

设计细节

地心引力织机是由一系列充满活力的彩色字符挂串组成的，横跨在整个博物馆的大堂上方，形成一种柔和的螺旋状薄纱表面。挂饰的表面扭曲，且盘旋在中庭，并以此对该空间做些改变，重新精心设计，来增强顾客流量与博物馆之间的交互作用。地心引力织机的字符挂串的色彩用来描绘传统的巴洛克式天花板的图案——即一个三维立体物体模糊成彩色的波涛，然后根据顾客的有利位置猛的集中成一个几何图形。

Stay Down Champion, Stay Down

流动装置

Project Location: USA
Architectural Design: SPORTS
Project Area: 84 sq. m.
Completion Year: 2010
Client: Woodbury University
Photography: Justin Harris
Building materials: Terracotta Tile

项目地点：美国
建筑设计：SPORTS 建筑公司
项目面积：84 平方米
竣工时间：2010 年
开 发 商：伍德伯里大学
摄　　影：Justin Harris
建筑材料：陶瓷砖

A Flowing Terracotta-Tile Structure 起伏的陶瓷砖建筑

Project Description

Stay Down Champion, Stay Down is a commissioned temporary installation for the Woodbury University Hollywood (WUHO) gallery, located along the busy and iconic Hollywood Blvd Walk of Fame. Considering this location and context, the installation focuses on the presence and activity of the ground plane. Comprised of a system of terracotta tiles, the floor of the gallery breaks away from the planar to compose a flowing and colorful interior landscape that guests can navigate through on foot. The bulging areas of this landscape are created by gradually elevating the handmade tiles on transparent acrylic supports. Heightened by vibrantly colorful lighting, the contrast of the analog tiles and digital surface creates a compelling visual and spatial experience that rivals the lure of the historic sidewalk beyond the gallery.

Design Details

The work of SPORTS explores the border between the everyday experience and the extraordinary or unexpected. By focusing on ideas of material, craft, and tradition, the design of SPORTS examines how these can both compliment and provoke contemporary notions of aesthetics and atmosphere.

项目介绍

该项目位于繁荣的好莱坞星光大道，是伍德伯里大学好莱坞（WUHO）画廊里一个临时装置。鉴于此种地理位置和前景，其设计侧重于它的外观和平面活动。该项目采用一系列陶瓷砖，画廊的地板脱离平面，构成一个流动的、丰富多彩的内部景观，客人可以通过步行浏览整个画廊。这一景观是在透明的亚克力材料的帮助下，将手工砖块设计为逐渐上升的形状。一个很有趣的充满活力的多彩照明装置艺术，用瓷砖搭建起一个起伏的流动效果。给人以视觉和空间上的奇特体验。

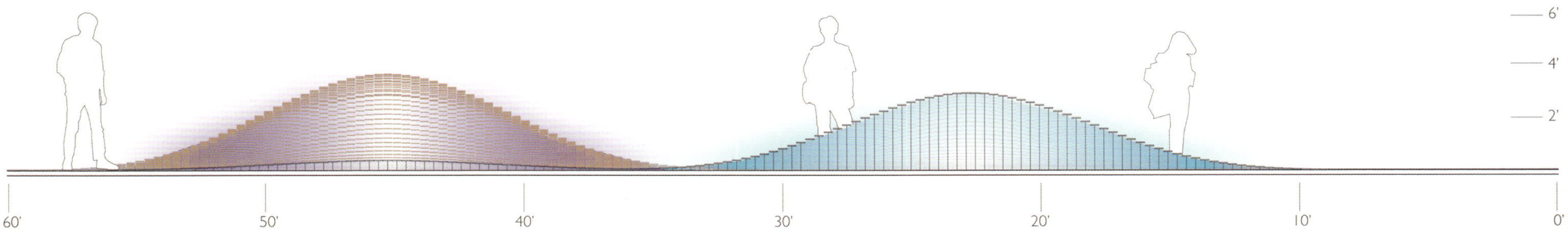

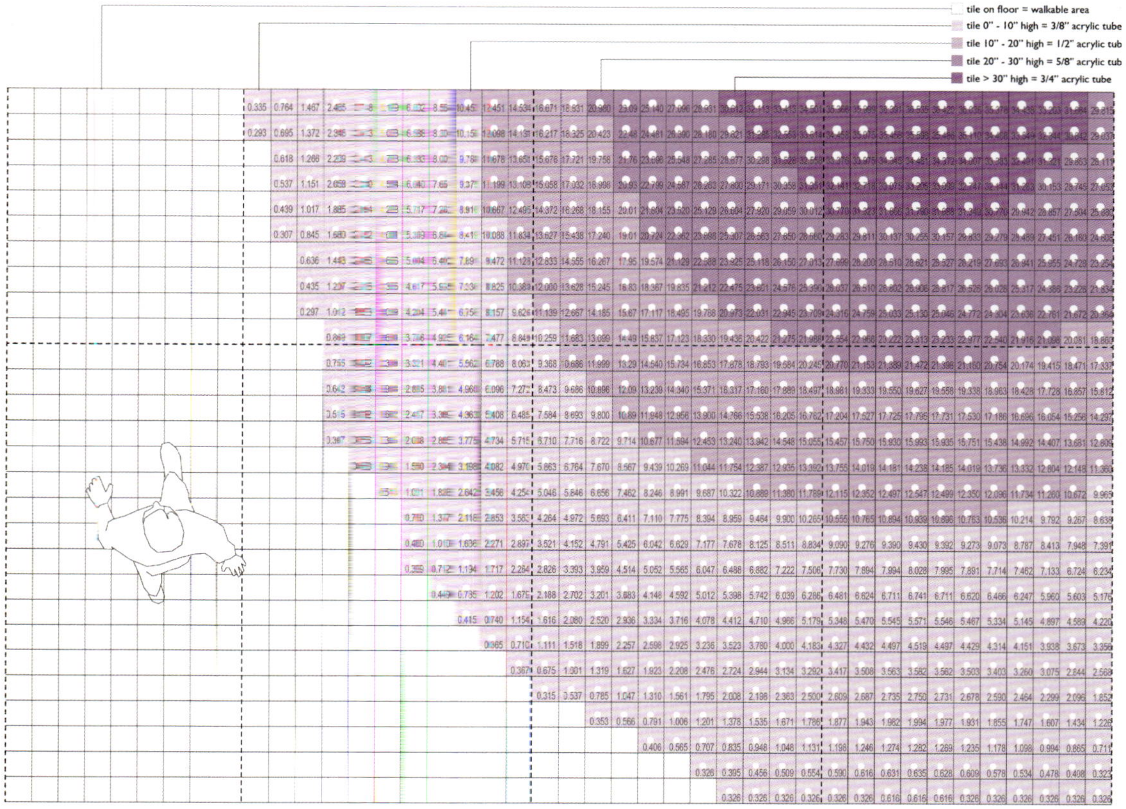

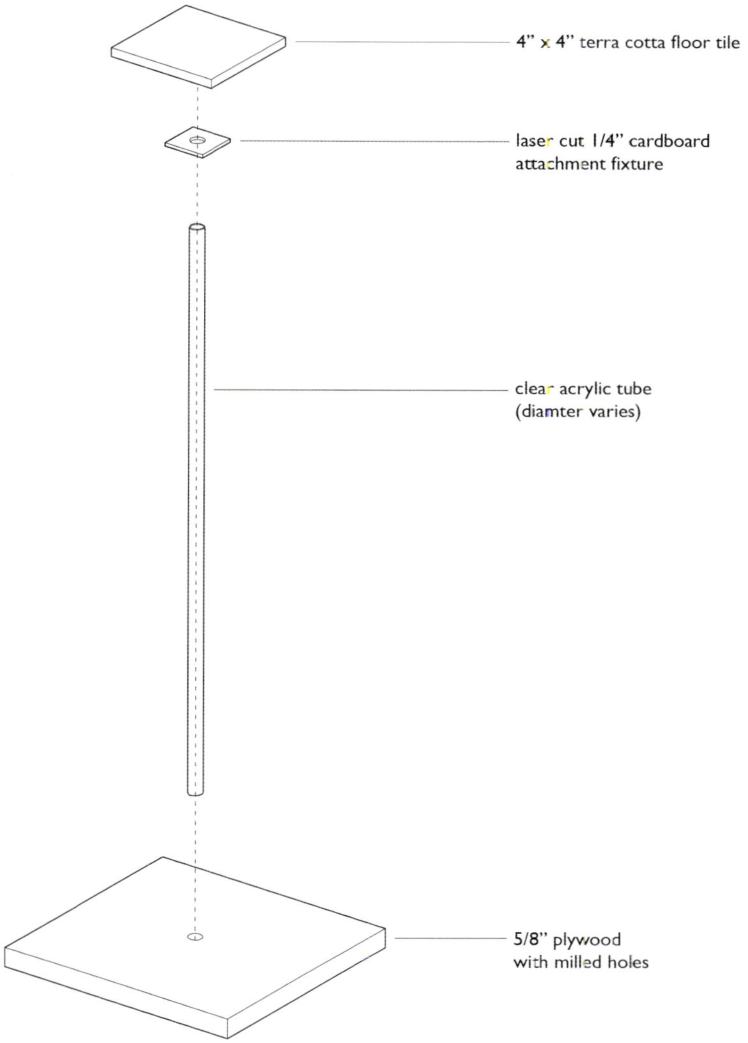

设计细节

SPORTS 建筑事务所探索了日常生活与非凡的生活之间的界线。设计师们将重点放在材料、工艺品和传统的选择上，证实了这些元素如何能称赞和挑衅现代美学和魅力。

Life will kill You

“生活让人疲惫”

Project Location: West Hollywood, California, USA
Architectural Design: SPORTS
Completion Year: 2010
Client: Revolve Clothing
Photography: Justin Harris
Building materials: Zip Tie

项目地点：美国加利福尼亚州西好莱坞
建筑设计：SPORTS 建筑事务所
竣工时间：2010 年
开 发 商：美国服饰 Revolve Clothing 零售网
摄　　影：Justin Harris
建筑材料：拉链带

Zip Ties Decoration 拉链带垂饰

Project Description

Life Will Kill You is a temporary installation for the Revolve Clothing showroom in West Hollywood. To stand in contrast to the high-fashion clothing of the boutique, an everyday industrial material—the zip tie, is aggregated to create a floating volume that nestles below an existing soffit. The design is intended to explore the edge between aggression and elegance through material sensibility, overall form, and visual effect.

Design Details

The cloud-like volume is created by a double-sided surface composed of over 100,000 zip ties. The exterior surface of the volume is an aggregation of longer, wider white zip ties while the interior is comprised of shorter and finer colored zip ties. The resulting bulging form offers ever-changing glimpses of blurred yet vivid color combinations as the zip ties layer on top of one another in the predominantly black and white store interior.

项目介绍

“生活让人疲惫”这一作品是为西好莱坞 Revolve Clothing 服装展厅设计的。为了与高品位奢侈品服装形成对比，设计师采用了一种日常工业材料，即拉链带，他们将许多拉链带连接在一起，在位于拱廊的下方，制造一个浮动的物体。该项目的目的是通过构造的敏感度和整体的形式，以及视觉效果来探索这个华丽之地。

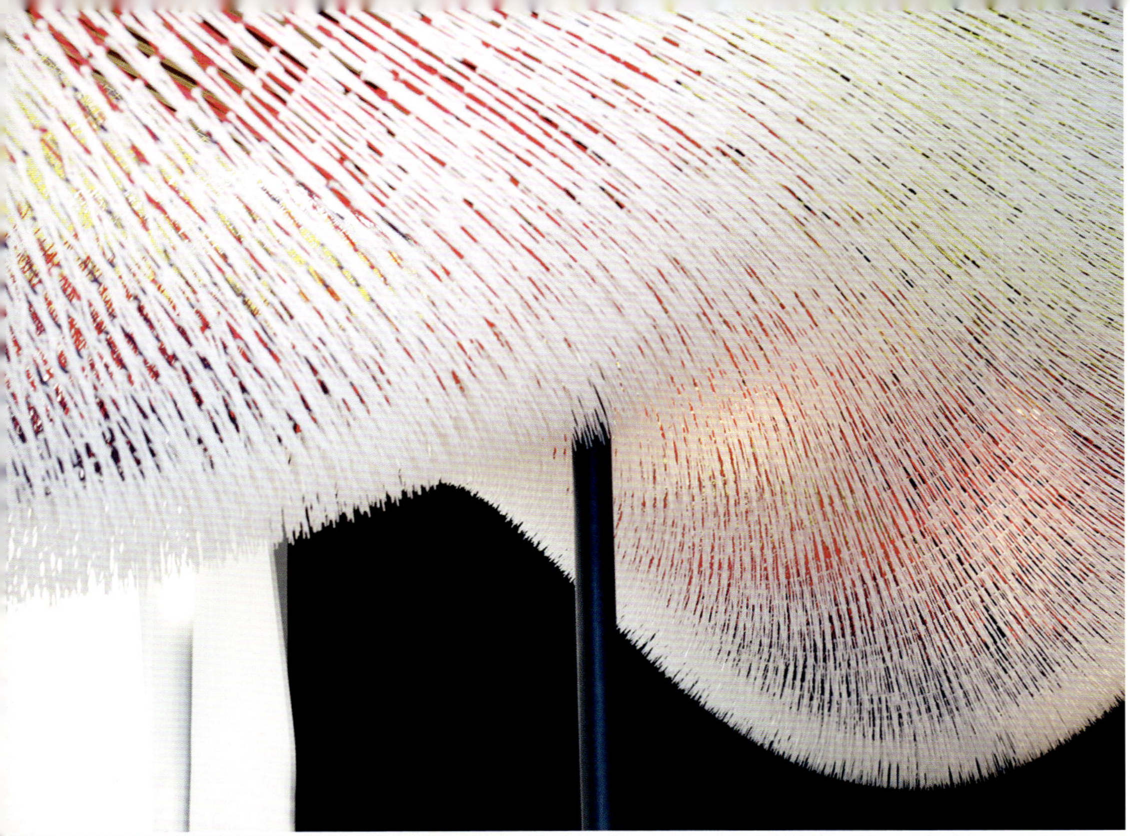

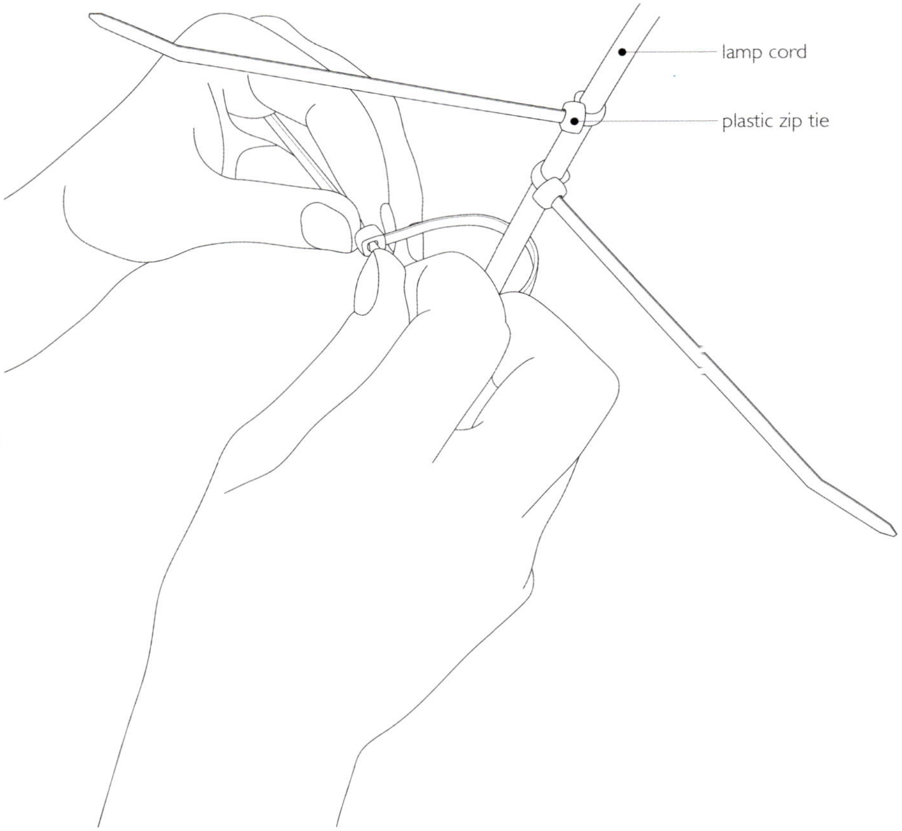

设计细节

这个似云一般的物体使用了 100，000 多个拉链带将它们连接成一张双面大网，悬挂在天花板上。中心部分垂落下来，并在人们头顶形成一个拱腹，它的表面是一些较长、较宽的白色拉链，里面是稍短、纤细一些的拉链。由于拉链层层叠加，并以黑色和白色作为底色，内部则使用较小的五颜六色的拉链，因此这个膨胀的装饰物生动的色彩组合给人一种千变万化的朦胧感。

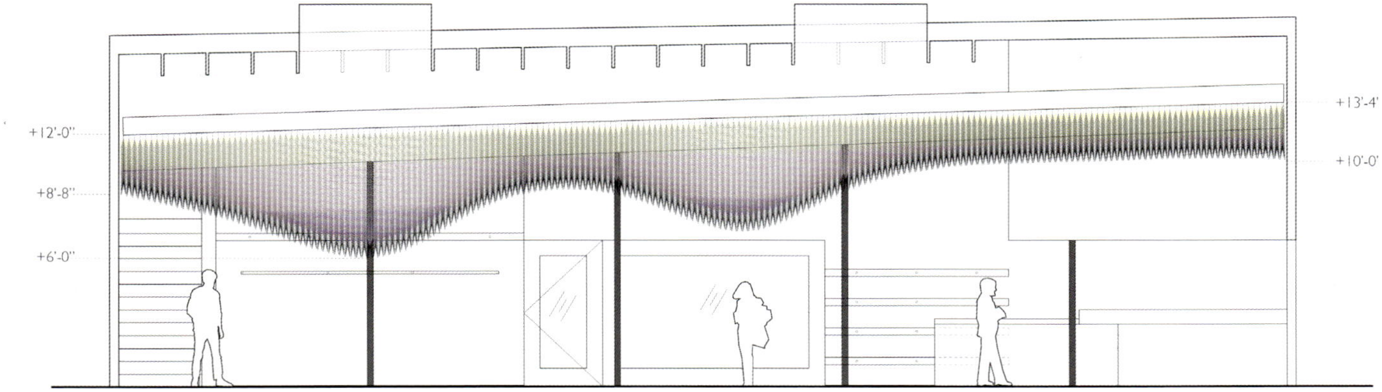

剖面图 /Sectional Drawing

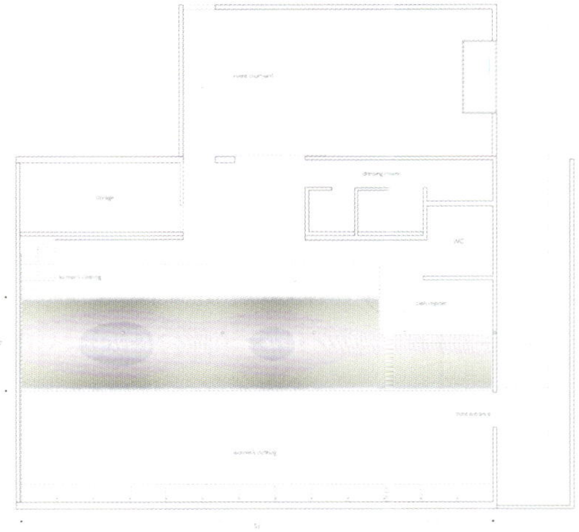
平面图 /Site Plan

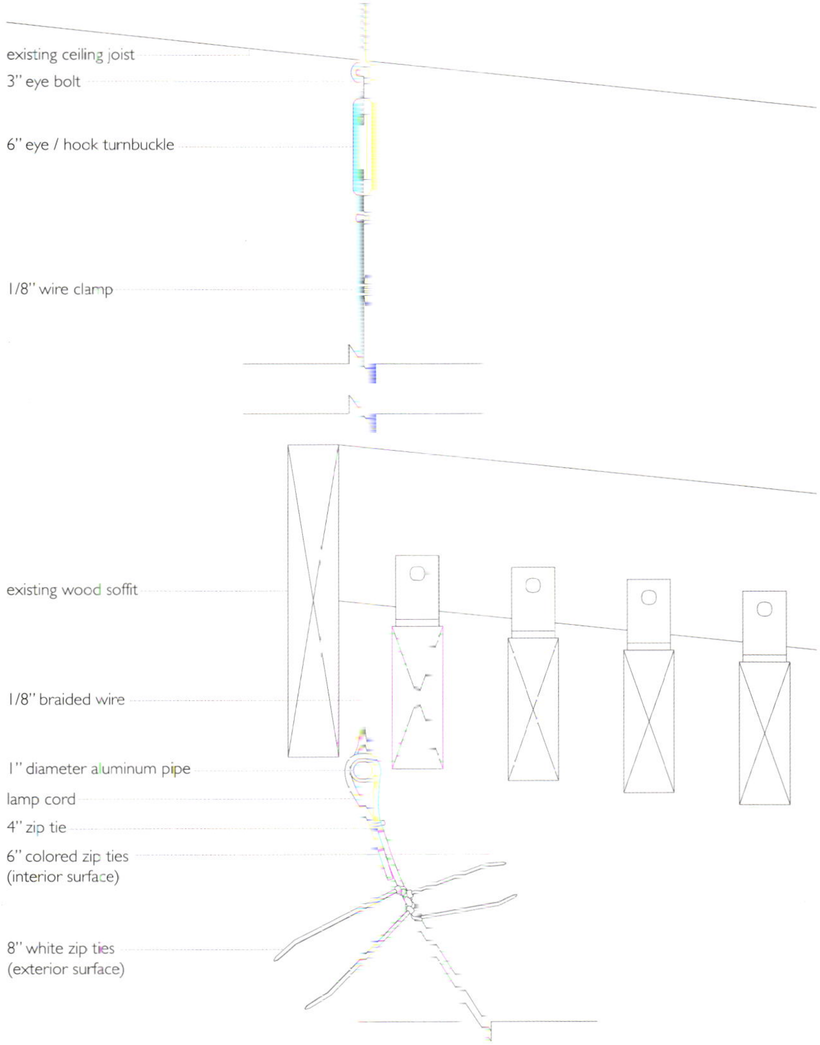

existing ceiling joist
3" eye bolt
6" eye / hook turnbuckle
1/8" wire clamp
existing wood soffit
1/8" braided wire
1" diameter aluminum pipe
lamp cord
4" zip tie
6" colored zip ties
(interior surface)
8" white zip ties
(exterior surface)

EXOtique

EXOtique 装饰

Project Location: USA
Architectural Design: PROJECTiONE
Completion Year: 2011
Client: The Institute for Digital Fabrication
Photography: PROJECTiONE
项目地点：美国
建筑设计：PROJECTiONE 建筑事务所
竣工时间：2011 年
开 发 商：数字化制造协会
摄　　影：PROJECTiONE 建筑事务所

Hexagonal Ceiling Design 六角形吊顶设计

Project Description

EXOtique is a quick design through fabrication installation at Ball State's College of Architecture. Through the use of computational tools (both software and hardware) the design through fabrication process can be drastically condensed. However the use of digital tools and design thinking for the purposes of fabrication rather than representation is not always at the forefront of designer's minds. This project was aimed at using tools purely as generators for fabrication without the need for representation. As designers and creative thinkers, we do not have to lean on these tools for the purposes of imagery; imagination can act as renderer, and understanding of materiality acts as logic in the physical world. Yet, in order to create the complex and infinitely different component systems, we are dependent on, and benefit from these digital processes.

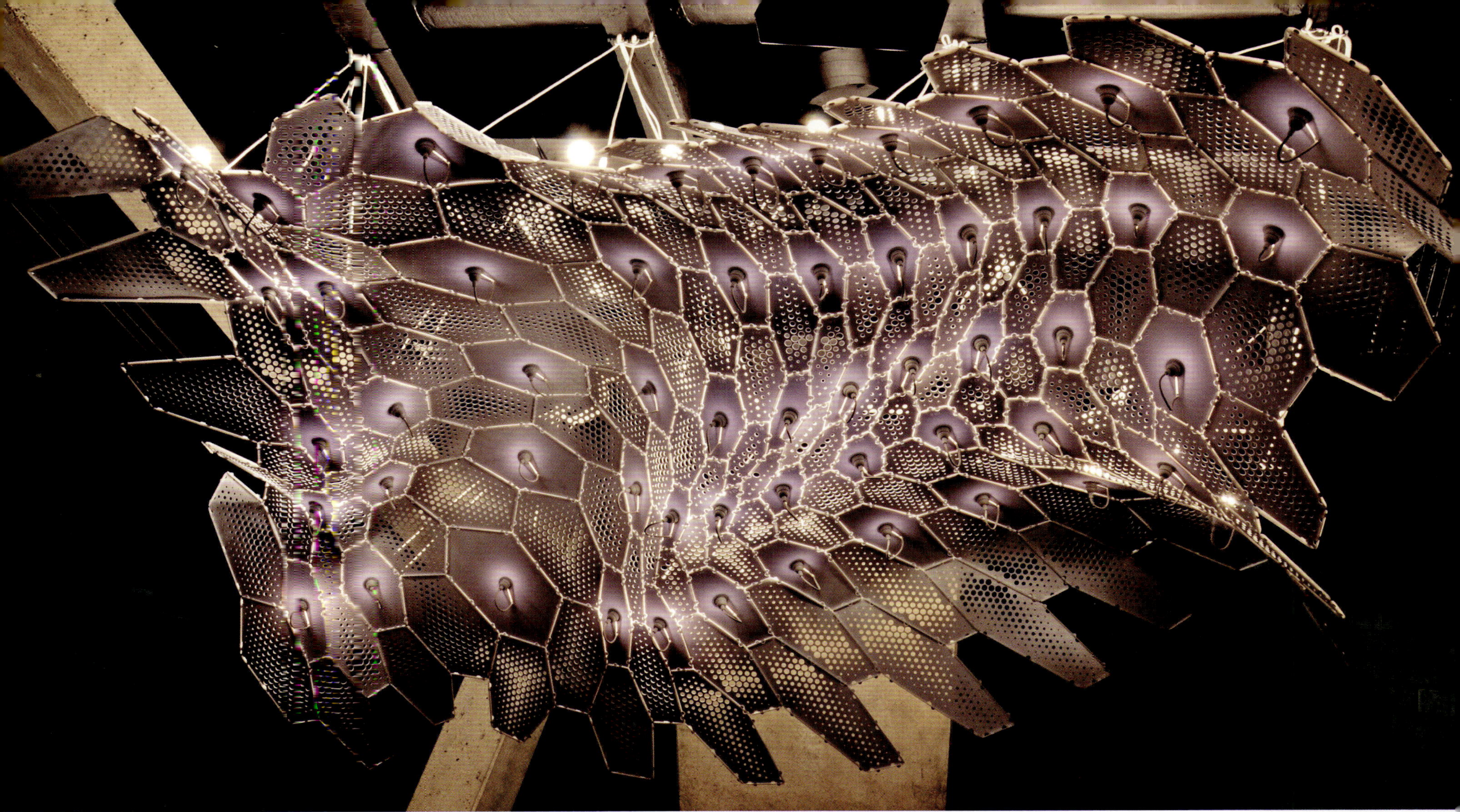

Design Details

Designers were intended to create a simple, hexagonally based, component system that would act as a lit "drop ceiling" for the space. All fabrication data was generated from a surface in Rhino, with all other processes accomplished through Grasshopper. The surface was triangulated and then associated into hexagonal groups. This was done with an understanding that the bending and folding could achieve non-planer geometry through the correct choice of material. The groups were unrolled and all other details were created. This includes joinery, labeling, patterning for distributed lighting, tolerance adjustments and other fabrication techniques (i.e. tabs for milling). Allowing flexibility within the computer to adjust tolerances made the feedback loop between digital and physical space easier and quicker. ExOtique combines an understanding of how materials react, and then intelligently embeds that into the design process.

B3.6
C3.3
B3.7
A3.5
C3.4
A3.6

项目介绍

EXOtique 是由鲍尔州立大学土木工程系在短期内制作安装的。该项目的制造过程利用了计算机工具（软件和硬件），大大缩短了工期。然而，数字化工具的利用和设计理念并不是设计者的最初想法。但最终该项目还是完全用工具替代了手工构图。作为创新设计师，他们并不是单纯依靠工具来设计图像，而是让想象变成渲染器，借助对物质在物理世界的逻辑理解，借助数字处理，创造出复杂的且与众不同的部件系统。

C3.3

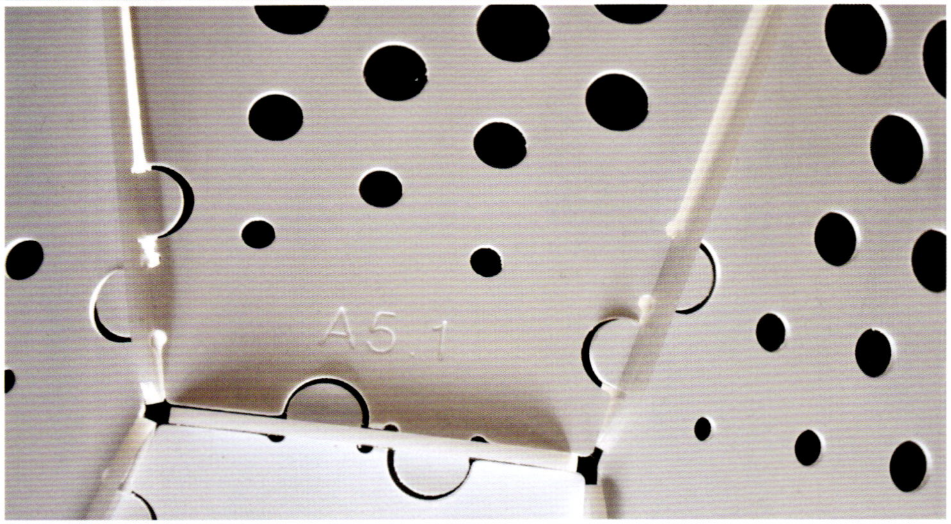
A5.1

Uniqa Vital Business Center

Uniqa 总部商务中心

Project Location: Budapest, Hungary
Architectural Design: Ferdinand Architects
Project Area: 32,500 sq. m.
Completion Year: 2009
Client: UNIQA-INVEST
Photography: Istvan Oravecz

项目地点：匈牙利布达佩斯
建筑设计：Ferdinand Architects 建筑事务所
项目面积：32,500 平方米
竣工时间：2009 年
开 发 商：UNIQA
摄　　影：Istvan Oravecz

A New Office Design 新办公空间设计

Project Description

The new head office is located close to the previous one at the junction of two important traffic arteries in District XIII, Lehel út and Róbert Károly körút. The office building constructed by property developer Raiffeisen Evolution in 22 months, named the Vital Business Centre, has net usable floor space of 18,000 square metres. "That means it is of a similar size to the Austrian foreign ministry".

Design Details

While designing the building the architect was guided by the principle of transparency. "Transparency is one of the cornerstones of UNIQA's business philosophy, and is conveyed in the Vital Business Centre by many glass surfaces and expansive rooms." A distinctive feature of the building is its LED facade, allowing colourful images and messages to be displayed on the street-facing side of the office building every evening. Although the idea is not new, it is unique in its scale: with some 80,000 pixels the Vital Business Centre currently boasts Europe's largest LED facade.

项目介绍

该公司的新总部离原先的办公室非常近，它位处十三区的两条重要的交通动脉上——Lehel út 及 Róbert Károly körút。该项目由地产开发商 Raiffeisen Evolution 修建，历时 22 个月，拥有 18， 000 平方米的网状实用面积，相当于整个奥地利外交部那么大。

Pénztár
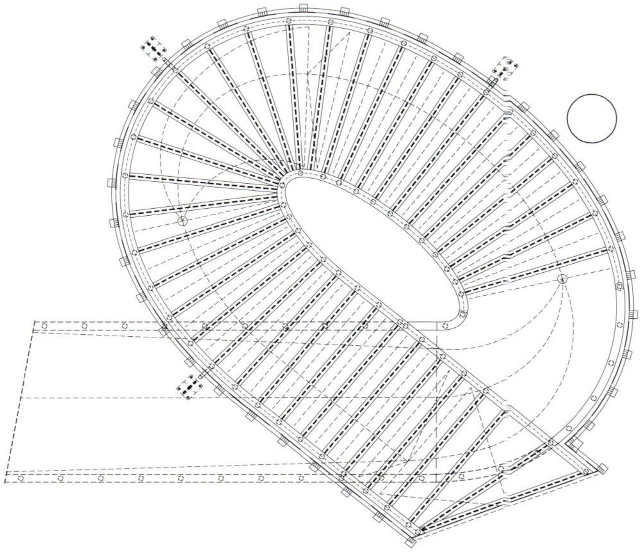
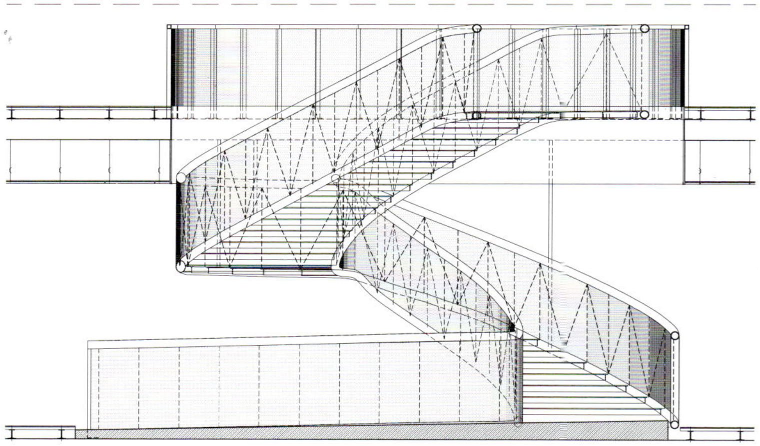

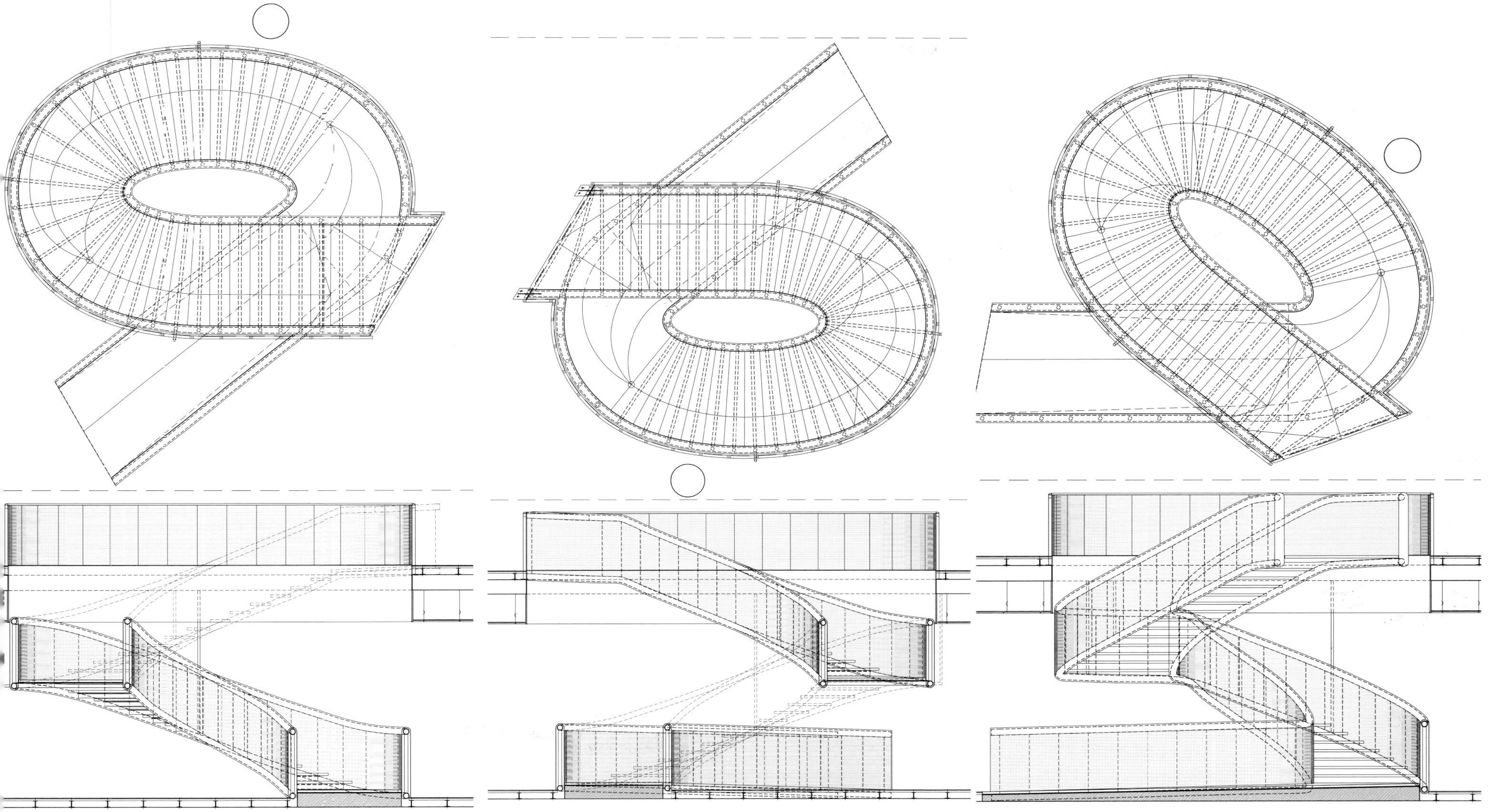

设计细节

"透明性"的原则贯穿整个设计的始终。"透明忙"是 UNIQA 商务经营哲学的基石，总部商务中心便以玻璃表面和宽敞的房间表达了这一理念。LED 外表面是该项目的特色之一。当夜幕降临的时候，面向街道的楼体上，会展示出色彩丰富的画面与文字。尽管这不是什么创新的设计，但它具有自身的独特性——它是整个欧洲最大的拥有 80 ，000 像素的 LED 灯建筑。

Out of the Strong Came Forth Sweetness

强壮来自甜蜜

Project Location: London, United Kingdom
Architectural Design: McChesney Architects
Completion Year: 2010
Client: Derwent London
Photography: Peter Cook
Building materials: Carbon Fibre, Leather

项目地点：英国伦敦
建筑设计：McChesney Architects 建筑事务所
竣工时间：2010 年
开 发 商：Derwent London
摄　　影：Peter Cook
建筑材料：碳纤维、皮革

An Oval Lounge in Angel Building 天使大厦里的椭圆形休息室

Project Description

Out of the Strong Came Forth Sweetness is a stunning new sculpture by Ian McChesney for the Angel Building in London. The shape of the piece was generated by allowing treacle to fall from a spoon—the resulting form is then inverted. The unit comprises an oval seating area from which extends a narrow twenty two meter high spar—that's over 5 double decker buses. The title is taken from the motto on the Lyles Black Treacle tin which, in turn is a reference to a story in the Old Testament. It is fabricated from carbon fibre which is both strong and very light enabling it to be incredibly slender. At the foot of the piece is a seating area upholstered in leather by designer Bill Amberg.

Design Details

The piece is made from carbon fiber which is both very strong and lightweight. The form was then designed in more detail by engineers Gurit using computer based 'Finite Element Analysis'. It was subsequently built on the Isle of Wight by AM Structures a fabricator with origins in boat building. The piece is 22 meters high and narrows to 100 mm diameter at the midpoint, and 25 mm at the tip. It has a structural steel core stopping at a height of 5 m from floor level. As the atrium was a main escape route in the event of fire, very limited amounts of combustible are permitted so there is less than 160 kg of carbon fiber used in the piece. It was installed in three sections, the base was one and the mast came in two halves.

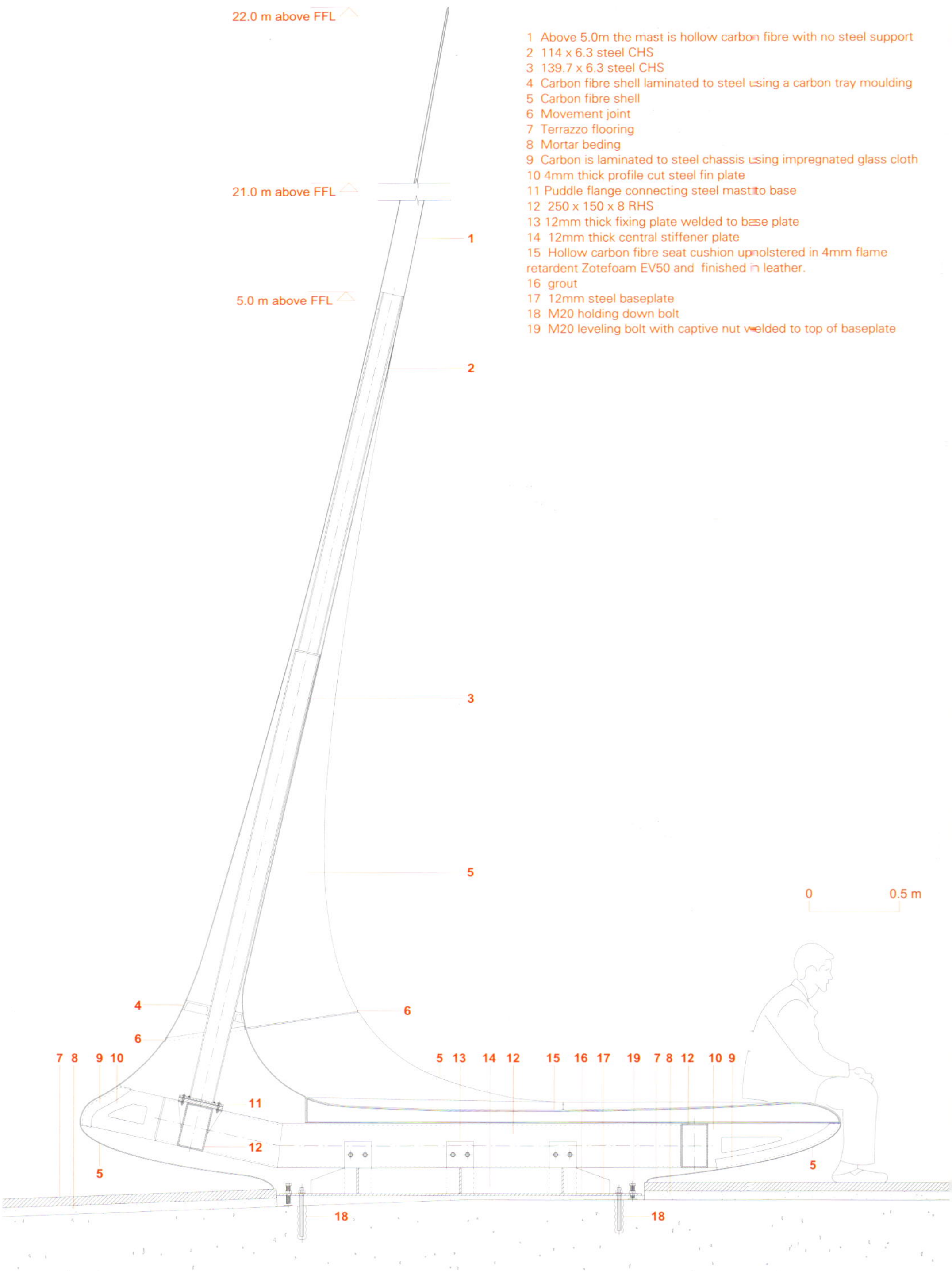

剖面图 /Sectional Drawing

项目介绍

“强壮来自甜蜜”是由 Ian McChesney 建筑事务所为伦敦天使大厦设计的一个惊人的新雕塑。该项目的外形设计就像从勺子里滴落的糖浆，因此最终形成了这一效果。该项目的设计包含从椭圆形休息区延伸出一个狭窄的 22 米高的晶石圆柱，超过了双层巴士的 5 倍高度。作品的名字“强壮来自甜蜜”就源自于莱尔斯黑糖浆。该装置包括一个 22 米长的延伸晶石和其下方的椭圆形休息区，该装置由碳纤维材料制成，因此非常坚固和轻盈，同时难以置信地纤细。休息区的坐垫是由著名设计师 Bill Amberg 设计的皮革软垫。

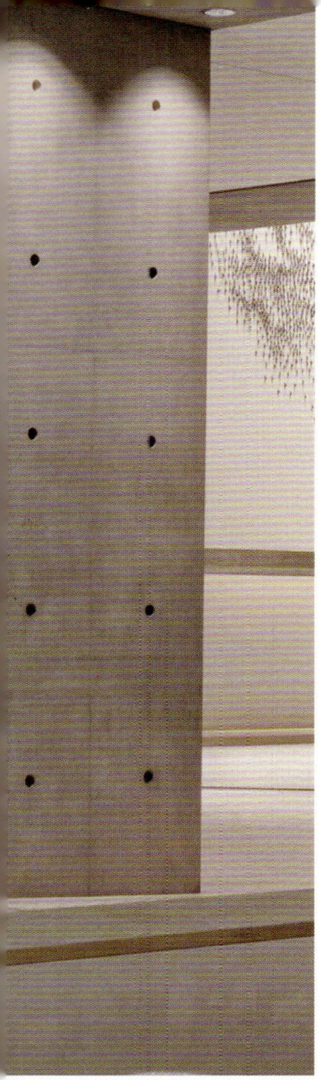
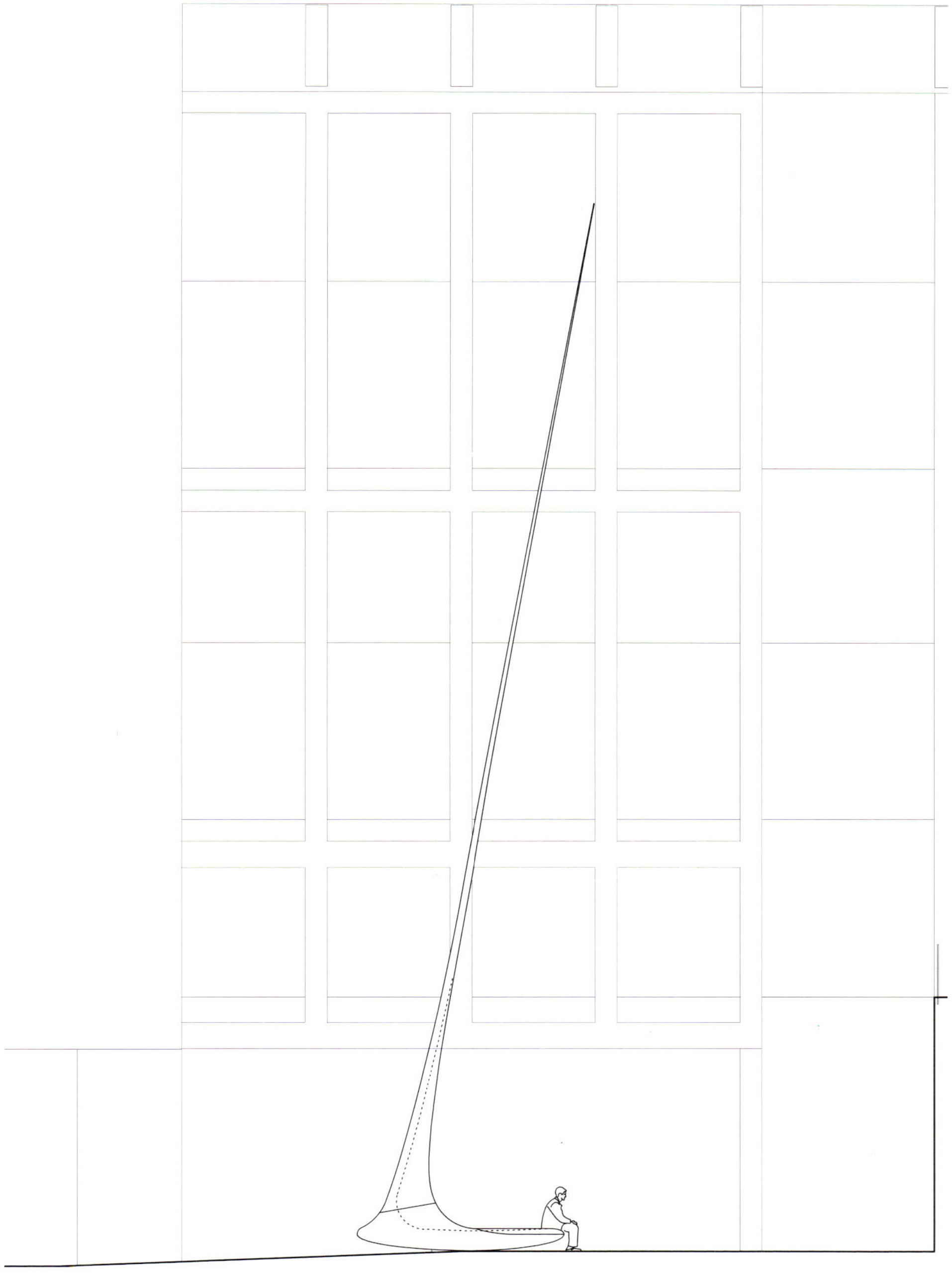

设计细节

该项目选用的是强度大，但重量轻的碳纤维材料。设计师与工程师经过“有限元分析”之后，对该项目进行了更为详细的设计，最后，该项目设计成 22 米长，受力直径缩小到 100 毫米，而顶端的直径仅为 25 毫米。在离地面 5 米高的地方有一个钢结构作为核心。由于中庭是火灾发生时的唯一逃生路线，可燃物使用的数量非常有限，所以该项目使用的碳纤维小于 160 公斤，并且分成 3 个部分，底部一部分，圆柱上的两面各一部分。

Alexandria

“亚历山大”

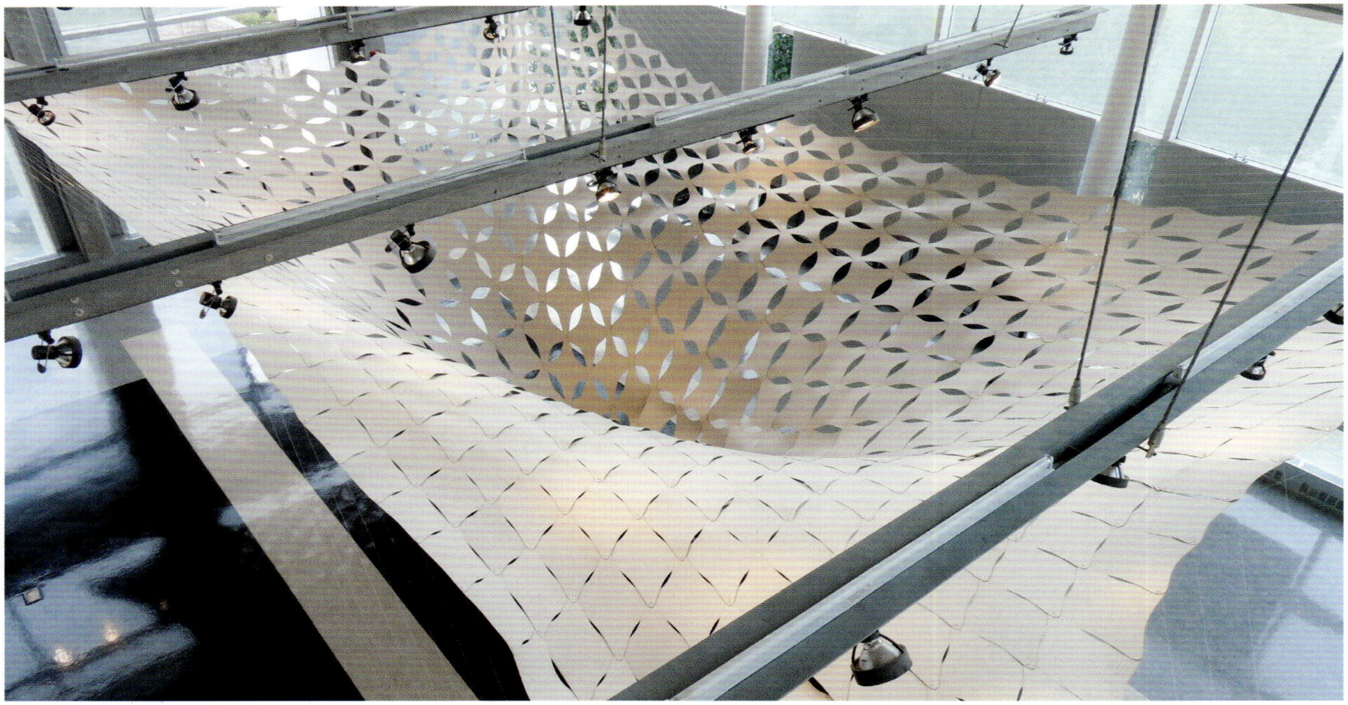

Project Location: Atlanta, Georgia, USA
Architectural Design: Anonymous Studio
Project Area: 93 sq.m.
Completion Year: 2010
Client: Southern Polytechnic State University

项目地点：美国佐治亚州亚特兰大市
建筑设计：Anonymous Studio 建筑事务所
项目面积：93 平方米
竣工时间：2010 年
开 发 商：南方州立理工大学

An Infrastructure in Southern Polytechnic State University
南方州立理工大学新建设施

Project Description

Alexandria is a project, designed and fabricated, by Anonymous Studio as an invited commission by Southern Polytechnic State University for the Architecture Gallery. With this installation, Anonymous Studio seeks to employ digital fabrication techniques and responsive conditions to create a physically and discursively interactive installation called Alexandria. Founders of the studio, Arseni Zaitsev and Jonathan Henry, utilize a simple conical figure emphasizing its place in the room and transitioning from a canopy system into an object in the space.

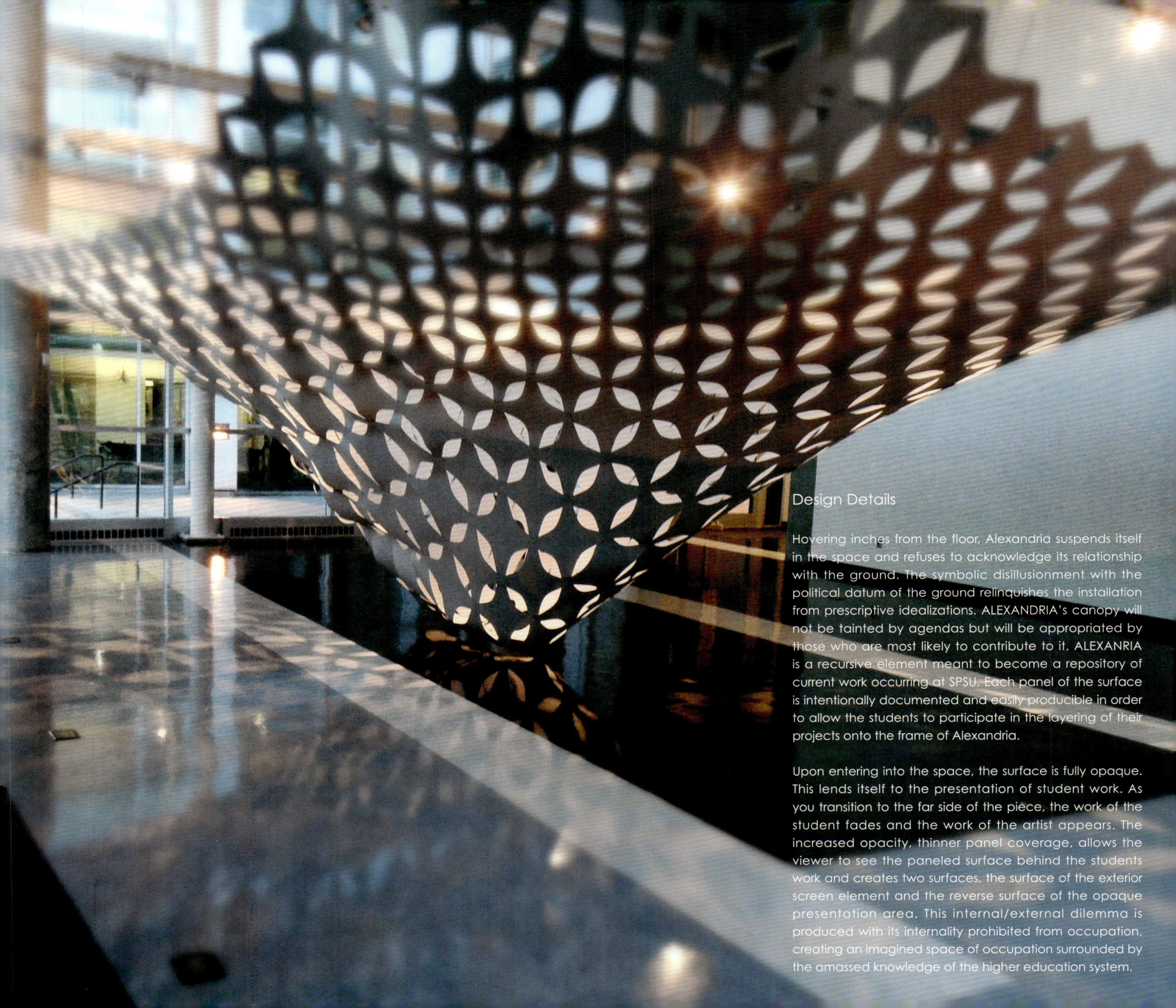

Design Details

Hovering inches from the floor, Alexandria suspends itself in the space and refuses to acknowledge its relationship with the ground. The symbolic disillusionment with the political datum of the ground relinquishes the installation from prescriptive idealizations. ALEXANDRIA's canopy will not be tainted by agendas but will be appropriated by those who are most likely to contribute to it. ALEXANRIA is a recursive element meant to become a repository of current work occurring at SPSU. Each panel of the surface is intentionally documented and easily producible in order to allow the students to participate in the layering of their projects onto the frame of Alexandria.

Upon entering into the space, the surface is fully opaque. This lends itself to the presentation of student work. As you transition to the far side of the piece, the work of the student fades and the work of the artist appears. The increased opacity, thinner panel coverage, allows the viewer to see the paneled surface behind the students work and creates two surfaces, the surface of the exterior screen element and the reverse surface of the opaque presentation area. This internal/external dilemma is produced with its internality prohibited from occupation, creating an imagined space of occupation surrounded by the amassed knowledge of the higher education system.

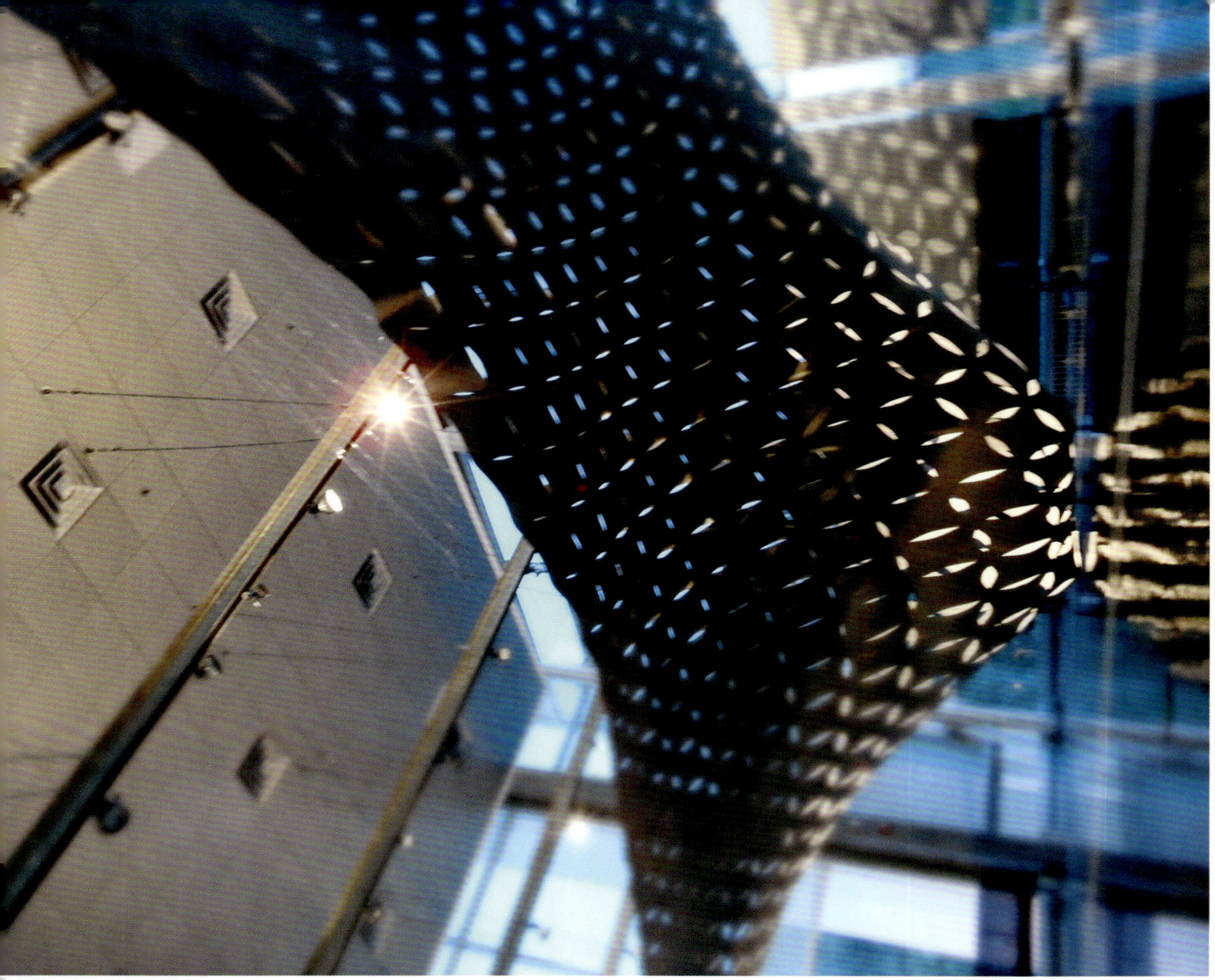

ALEXANDR

项目介绍

“亚历山大”这个项目是 Anonymous Studio 建筑事务所受南方州立理工大学之邀，为建筑画廊设计制造的。该项目的设计旨在运用数字化制造技术和反应条件，来创造一个身体和语言上的交互式装置，并命名为“亚历山大”。该设计公司的创始人 Arseni Zaitsev 和 Jonathan Henry，利用一个简单的锥形图，强调其在房间所放置的位置，仿佛从遮篷转换成空间中的一个物体。

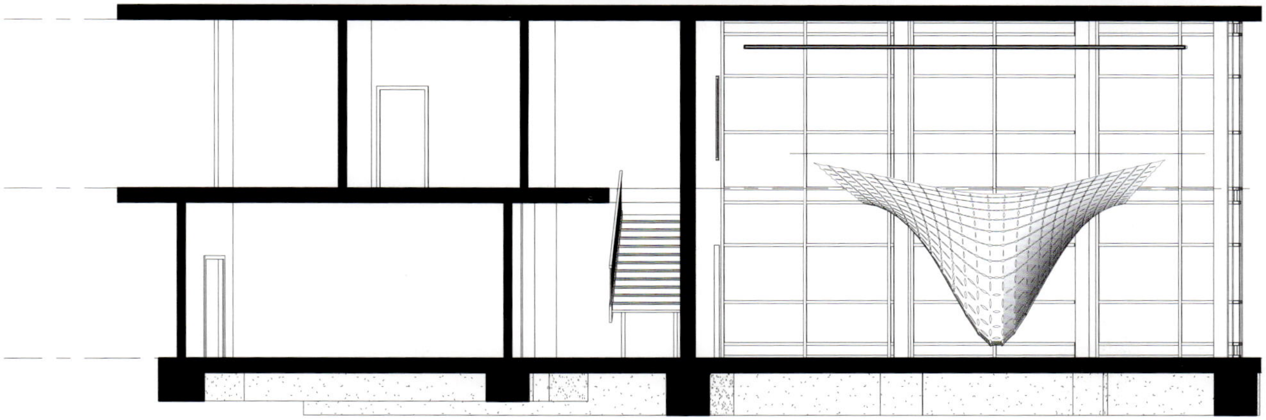

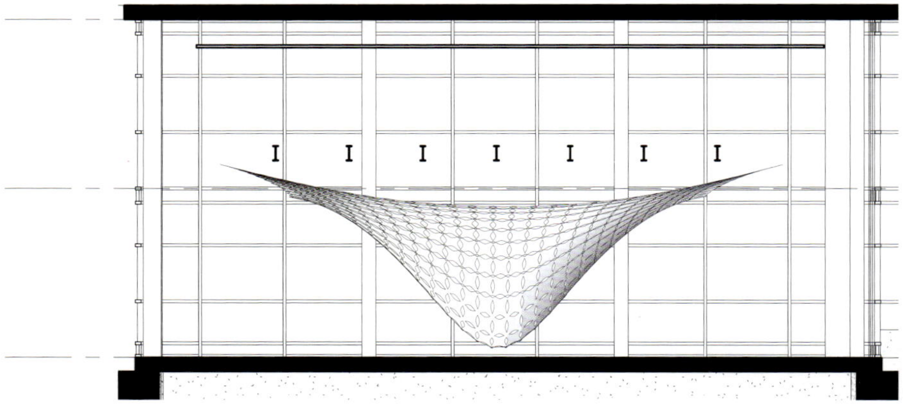

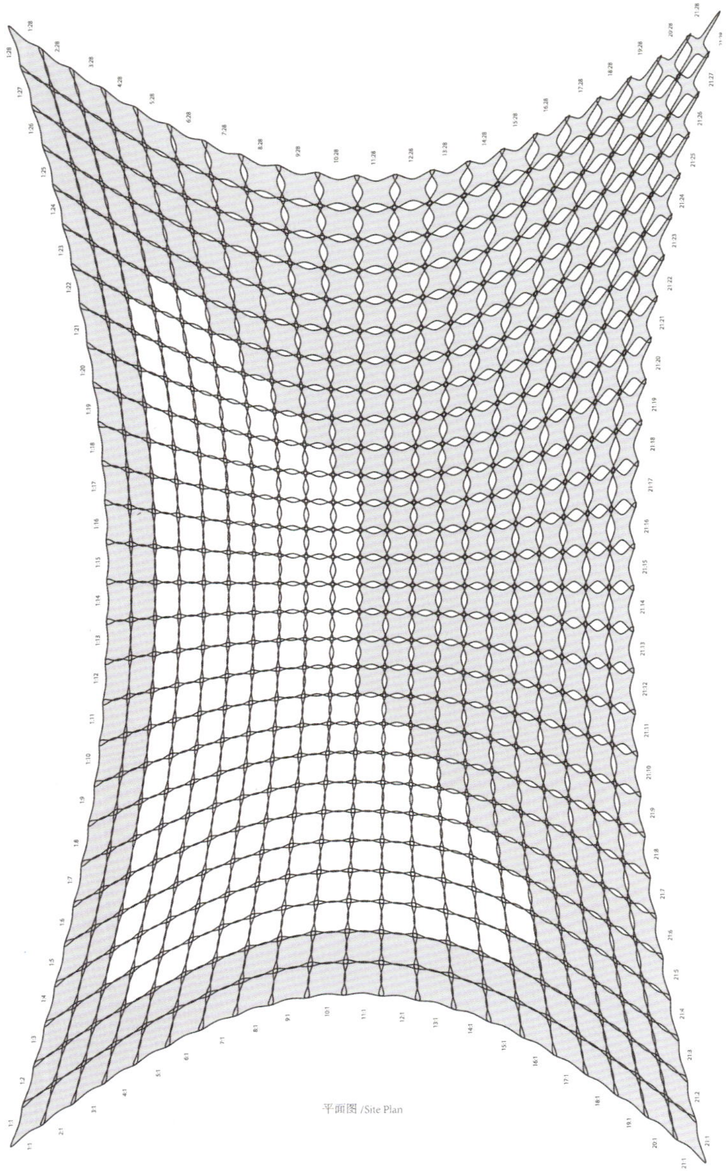

平面图 /Site Plan

设计细节

“亚历山大”盘旋在半空中离地面几英尺处，它悬浮在空中，没有与地面直接接触。由于地面上许多设计基准，让设计师放弃了将它设计成约定俗成的装置。“亚历山大”的遮篷不会被日常事物所污染，但仍然会被一些最有可能做学术的人使用。“亚历山大”是一个循环装置，开发商想让它成为南方州立理工大学里一个记录当前工作的宝库。该建筑每个面板的表面都作了记录，并可轻易延长，目的是允许学生参加在“亚历山大”框架上的分层活动。

进入这个空间，表面完全是不透明的。这里适合学生做工作陈述。如果你再走近一点，学生的工作开始变得模糊，艺术家的工作呈现出来。设计师增加了一些不透明度和较薄的覆盖面板，允许观众从学生作品的后面就能看到镶板表面，同时制造出双层效果，一面是外面屏幕的表面，另一面是不透明展示区的反面。这种内部与外部困境的产生来自于禁止占领的内在素质，创造一个用高等教育体系积累的知识所包围的想象空间。

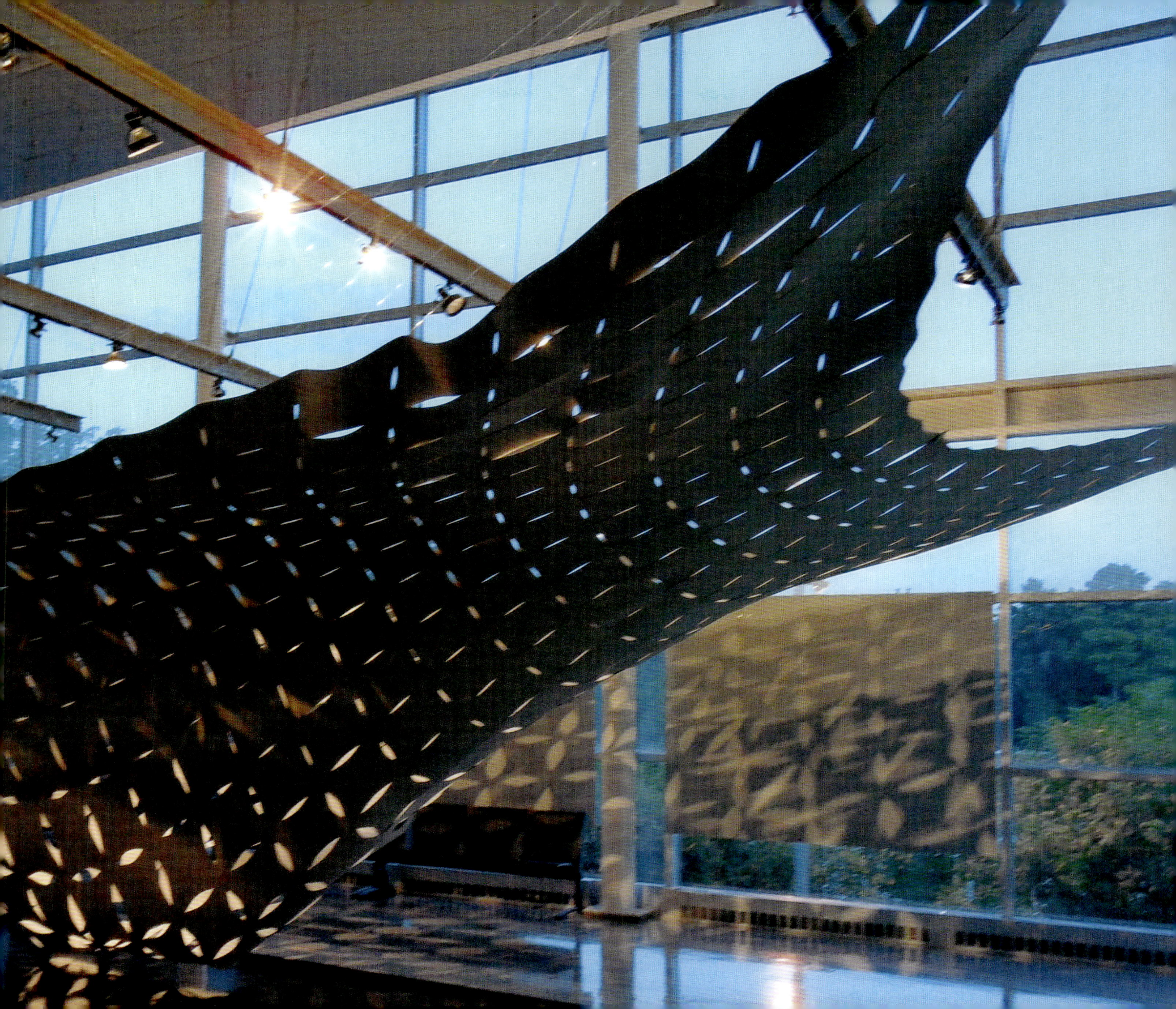

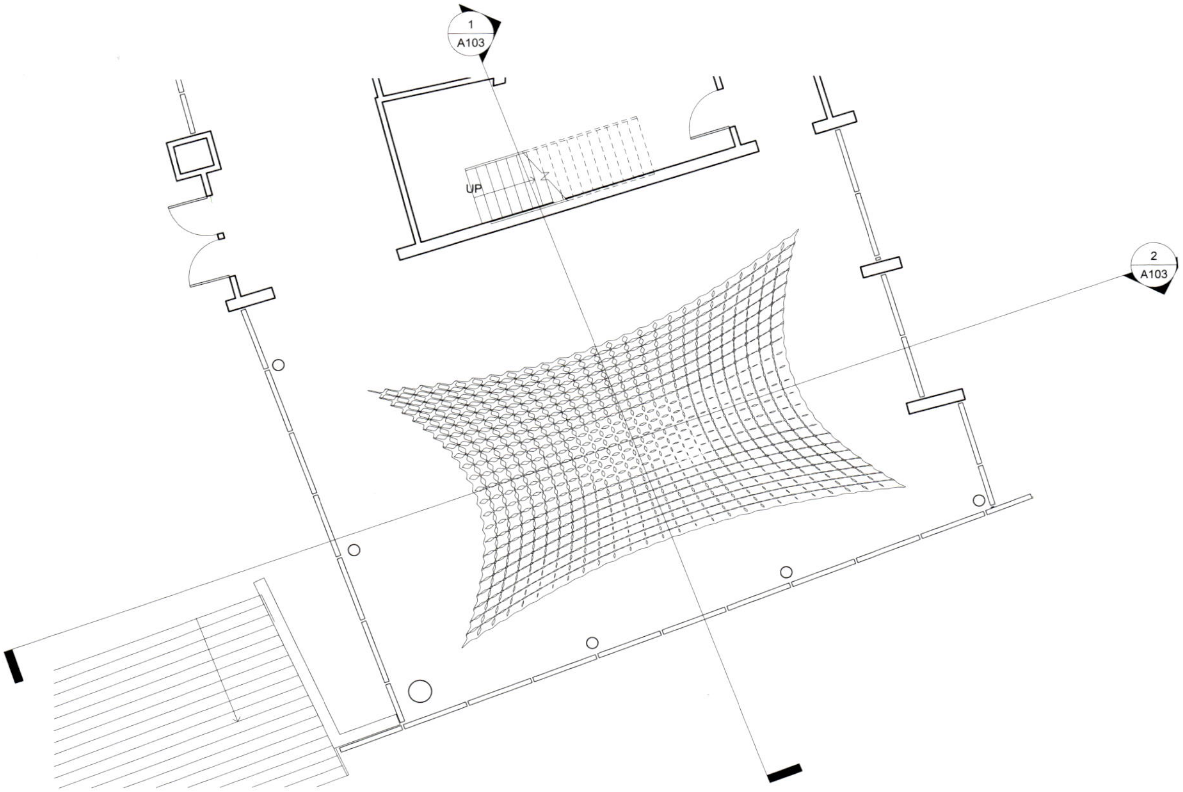

1
A103
UP
2
A103

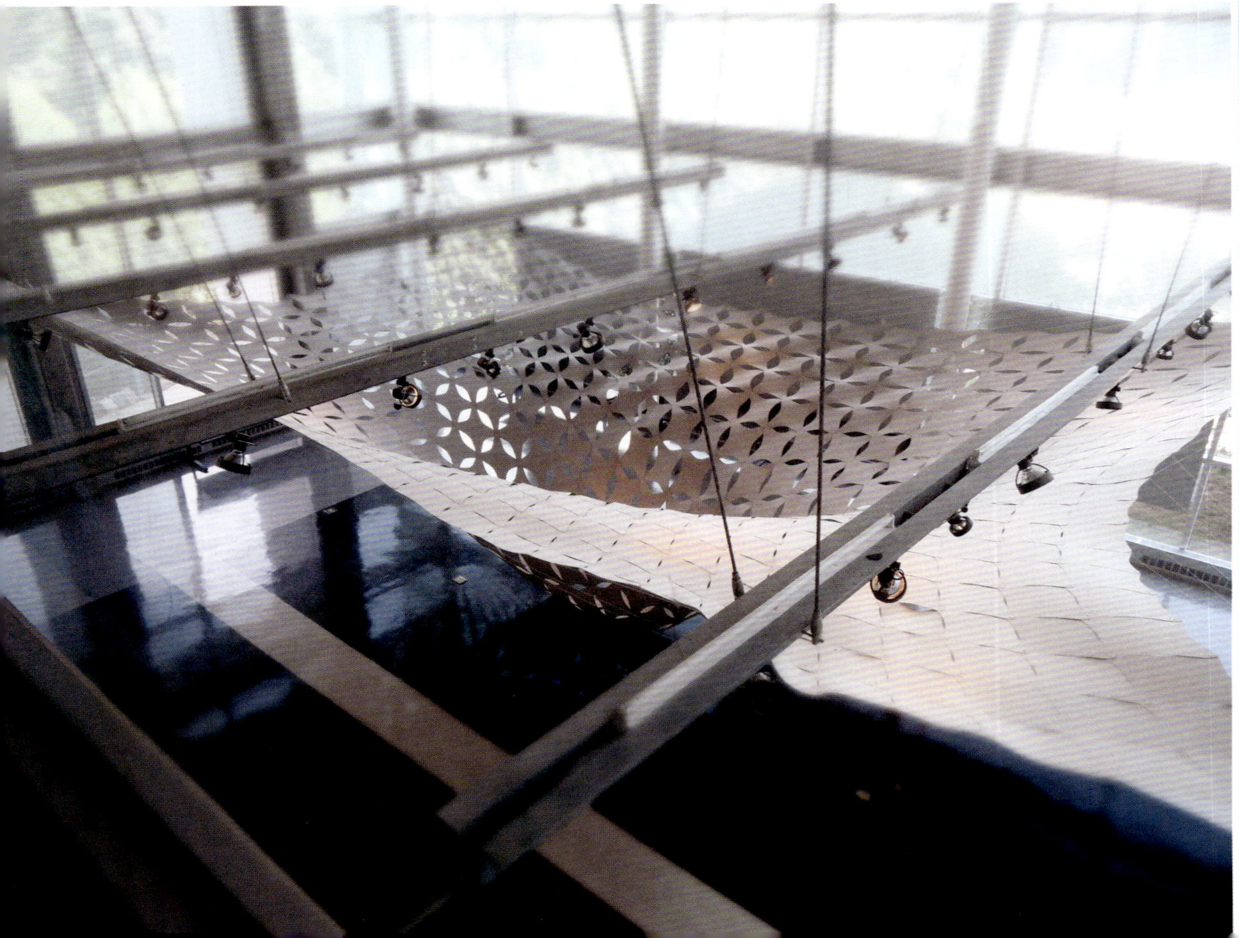

Field's Hi'Street

Field's Hi' 大街

Project Location: Denmark
Architectural Design: Rosan Bosch
Completion Year: 2010
Photography: Rosan Bosch & Rune Fjcrd

项目地点：丹麦
建筑设计：Rosan Bosch 建筑事务所
竣工时间：2010 年
摄　　影：Rosan Bosch & Rune Fjord

A New Entertainment Space in Shopping Mall 丹麦购物商场的新装饰

Project Description

Who said that it is boring to do your shopping in a shopping mall? The anonymous arcade at Field's shopping mall has been transformed into a dynamic Hi' Street that makes it an experience to replenish the wardrobe, and creates a clear contrast to the traditional shopping environment seen in other shopping malls. From being an overlooked part of the centre, the arcade has now become a special destination—a desirable area that now is the first place rather than the last one visits in Field's.

MAKE UP STORE
ROYAL
d'luxe

Design Details

Before, customer behavior towards the arcade was marked by a relatively modest influx of people than the rest of Field's due to its somewhat secluded location in a corner of the mall. Now, Field's new Hi' Street stands out as a unique destination with intersecting, green structures, challenging conversation furniture, inviting lounge areas, sound works and a street art wall. Hi'Street is a mixture of fashion, art and design—an alternative city space that makes shopping a sensuous and stimulating experience. This is where you see, but also where you are seen by others. Field's Hi' Street has gained a strong identity and clear signaling effect as well as the character of being a hidden gem that everybody has to see—an exclusive and inspiring oasis with a creative atmosphere and curious experiences that use the spatial qualities of the arcade to full, giving it a unique character. Aesthetically, the structure serves as a visual attraction, adding dynamic and surprising qualities to the space. It makes Field's customers curious and inclines them to take a closer look.

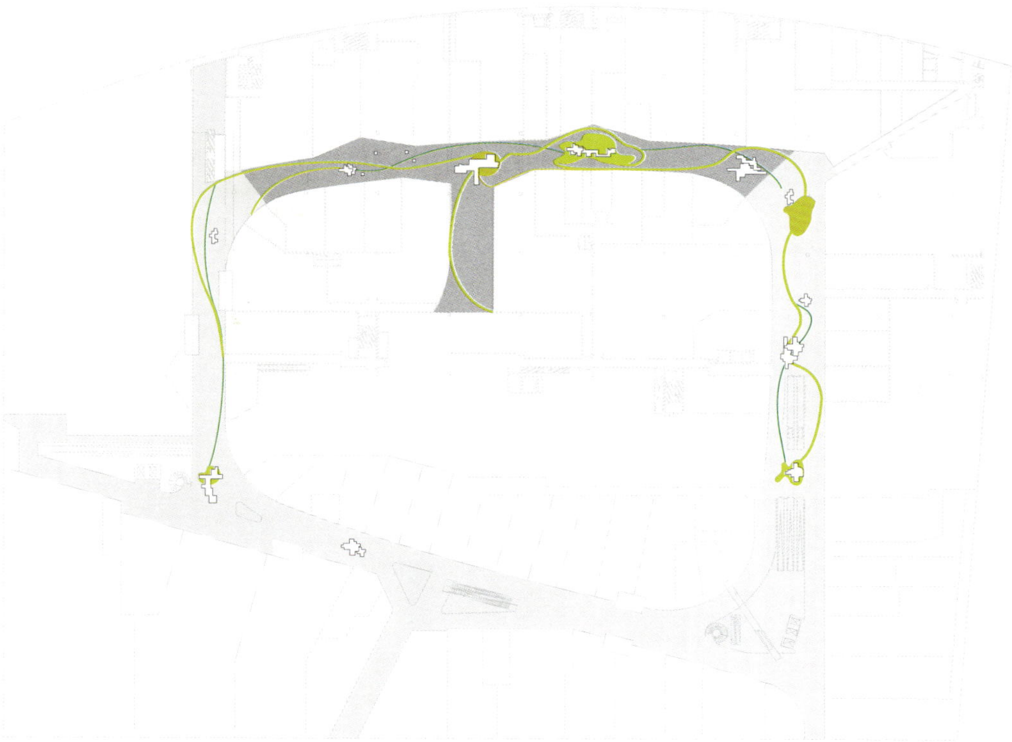

全景平面图 /Panoramic View Plan

inatricot

项目介绍

谁说在商场购物是枯燥的呢？位于 Field 购物中心的匿名拱廊形街道已将它转换成一个充满活力的 Hi' Street，它可当作衣柜，并与其他商场看到的传统购物环境形成明显的对比。该项目从购物中心被忽视的部分，变为该商场特殊的目的地，如今它已成为该商场首个被顾客光顾的地方，而不是最后一个被拜访的场所。

设计细节

在此之前，拱廊位于商场较隐蔽的角落，顾客对拱廊的青睐仍没有对商场其他区域那么浓厚，人流量也相对较少。而现在，Field 商场的新标志性建筑——Hi' Street，脱颖而出，交叉的绿色结构，富有挑战性的家具可供游客在此洽谈，富有魅力的休息区、健全的设计和一面街边艺术墙，成了顾客们唯一的目的地。Hi' Street 的设计融合了时尚、艺术和设计，将它打造成另一种城市空间，给予购物一种感性与刺激的体验。Field's Hi ' Street 大街已获得强烈的特征和明确的信号效果，好似隐藏的宝石，但人人可见。具有创造性的氛围和奇特感受的唯一且鼓舞人心的绿洲，用拱廊的空间特质来设计，赋予它一个独一无二的特征。从美学角度来说，该项目具有强烈的视角吸引力，为空间增添了动态感并产生了惊人的效果。这使得顾客对它充满了好奇，从而停下他们的脚步俯身看个仔细。

Drop-In Distraction

垂链装饰

Project Location: USA
Architectural Design: Ball-Nogues Stucio
Completion Year: 2009
Photography: Joshua White
Building materials: Metal bead chain, Aluminum Sheets

项目地点：美国
建筑设计：Ball-Nogues Studio 设计公司
竣工时间：2009 年
摄　　影：Joshua White
建筑材料：金属珠链、铝板

The Art of Ceiling 天花板上的艺术

Project Description

This hanging sculpture for the Los Angeles County Building and Safety Permit Office uses approximately two thousand individual lengths of metallic bead chain hanging under self-weight to form a matrix of catenary curves. A combination of sculptural artwork and modular ceiling system, the chains span between custom perforated aluminum panels fitted within the existing acoustical ceiling grid. Each chain is in precise relation to its neighbors to yield an array that is more a diaphanous metallic vapor than a discrete solid object.

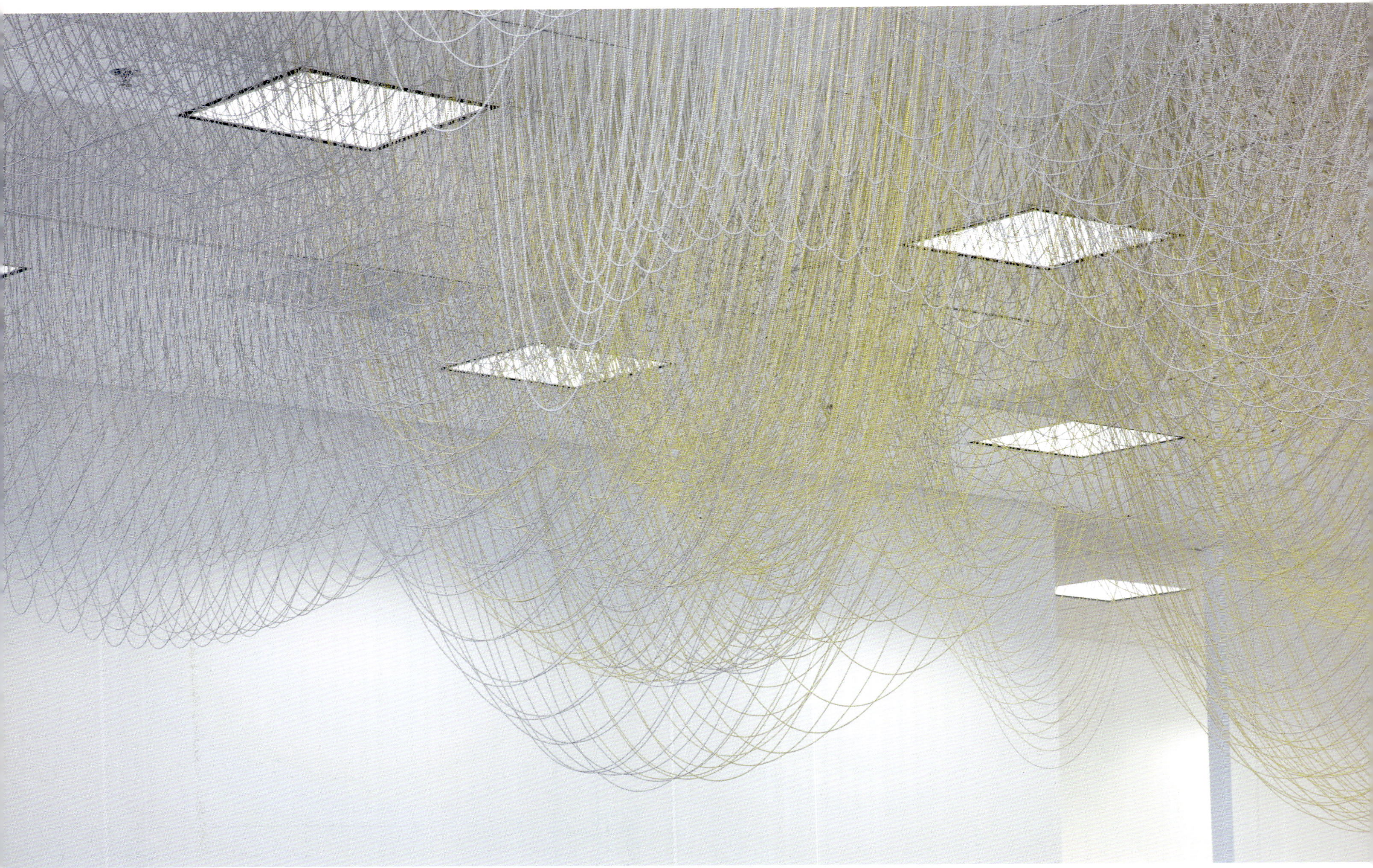

项目介绍

该项目是为洛杉矶的县建设和安全生产许可证办公室设计的装饰物，设计师采用了大概 2,000 多个金属珠链形成弯曲的悬垂线。设计师将雕塑艺术品和模块化天花板相结合，穿孔铝板之间的珠链填充了天花板的网格。每条珠链都与周边的链子相连，形成一列透明的金属蒸汽，而不是一个离散的固定物体。

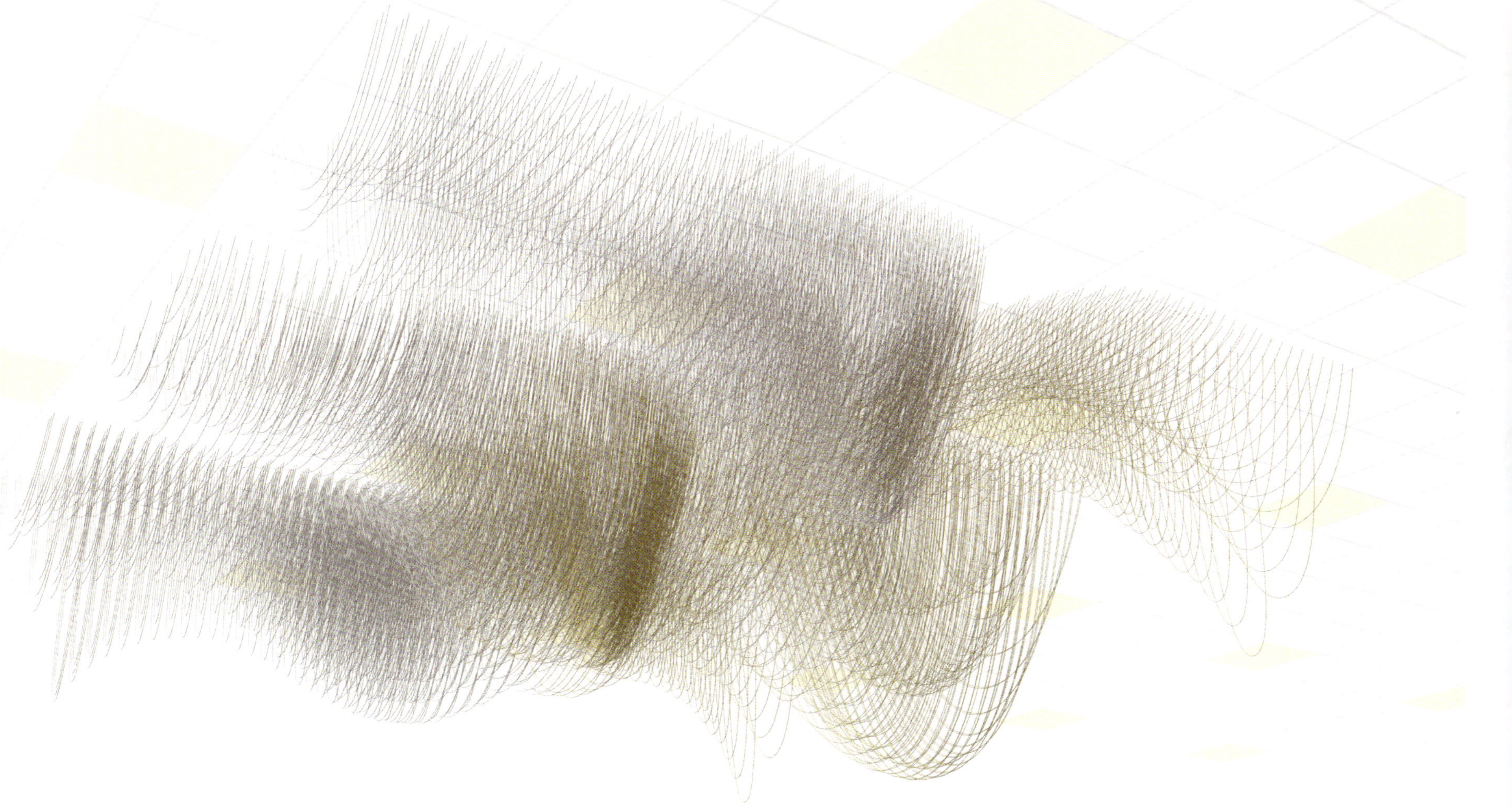

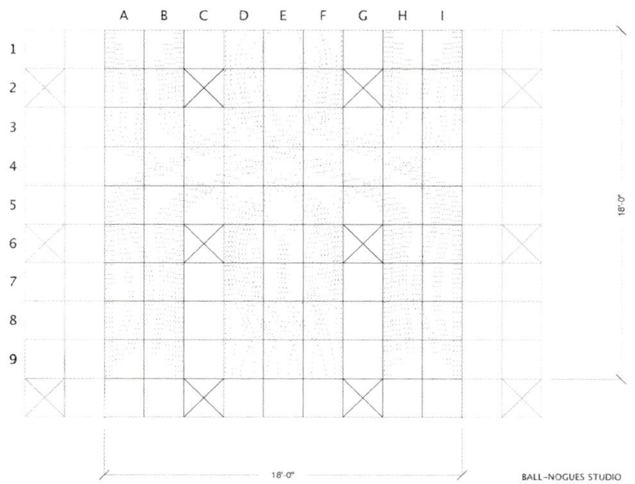

BALL-NOGUES STUDIO
PERMIT OFFICE
SCHEME K-09
12/17/2008

SUPPLIES

Canopy

遮篷

Project Location: UK
Architectural Design: United Visual Artists
Completion Year: 2011
Project Area: 270 sq. m.
Building materials: Module

项目地点：英国
建筑设计：联合视觉艺术团体
竣工时间：2011 年
项目面积：270 平方米
建筑材料：模块

A Shining Embellishment At Entrance 入口处的华丽点缀

Project Description

During the day, natural light reflects and refracts through apertures in the modules to the pavement below; at night, the canopy comes to life. Particles of light are born, move around their environment, and then die. Although the piece is hard, angular and mechanical, the overall impact is softer and more organic—reminiscent of a forest canopy, with dappled light modulated by passing weather systems.

Convention
Centre
North/South
citi
Public Parking

Design Details

Designers set out to create a sculpture that simultaneously evoked man-made rationality, natural irregularity, and blured the distinction between the two. The result was Canopy, a work consisting of over 8,000 identical polygonal modules which together form a 90 meters long organic mesh suspended above the sidewalk. Canopy is a regimented grid whose modules recall crystalline materials, the cells of leaves, and the shape of the maple leaf itself. The modules grow outwards from seed locations to colonize the grid; subsequently not every cell is occupied by a module.

项目介绍

白天，阳光透过模块上的缝隙照射和反射在下面的人行道上。到了夜晚，遮篷便恢复了生命。虽然每一片材料都是坚硬的、多边形的、机械的，但总体上给人一种柔和与有机的感觉，让人联想到一片森林的遮篷，根据不同的阳光映照出斑驳的光线。

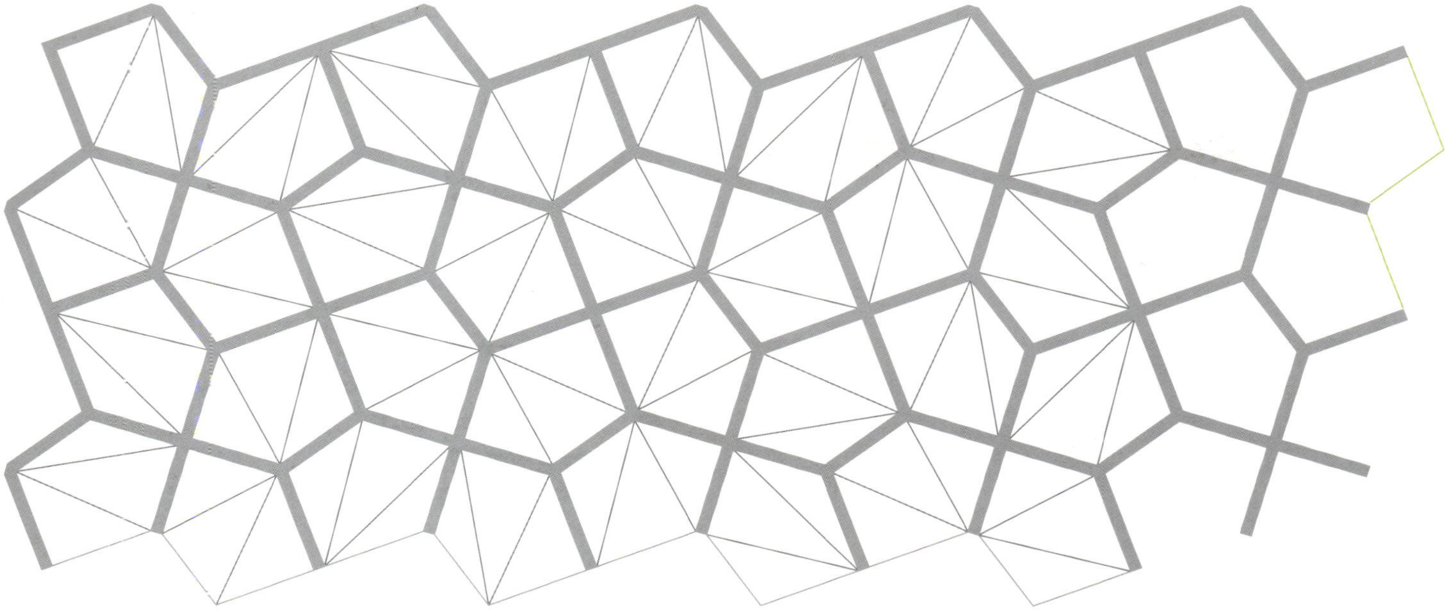

设计细节

这一雕塑的设计，在实现美感的同时，也激发了人为理性、自然的不规律性，并淡化二者之间的区别。该项目是由 8，000 多个相同的多边形模块组成的，形成一个 90 米长的有机网状，悬浮在过道上。设计师将遮篷设计成网格形，模块采用像枫树树叶一样的细胞组合而成的晶体材料制作，向外设计，形成网状。

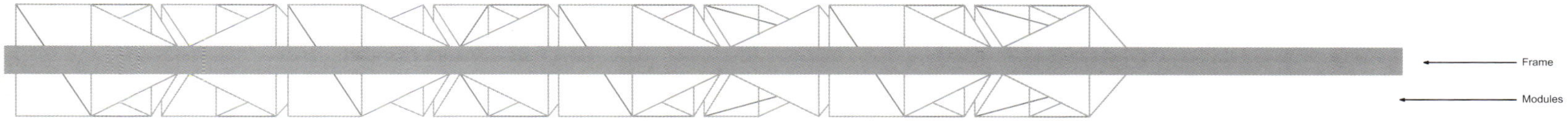

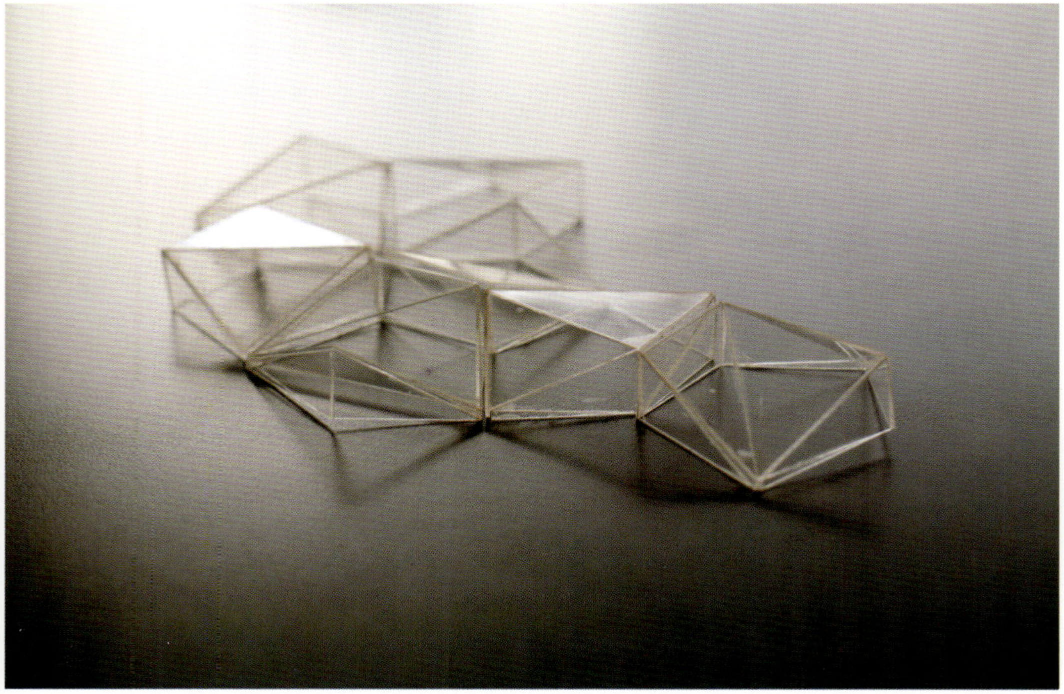

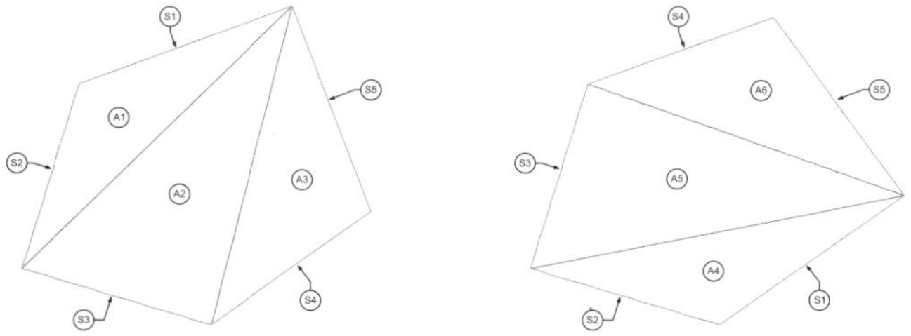

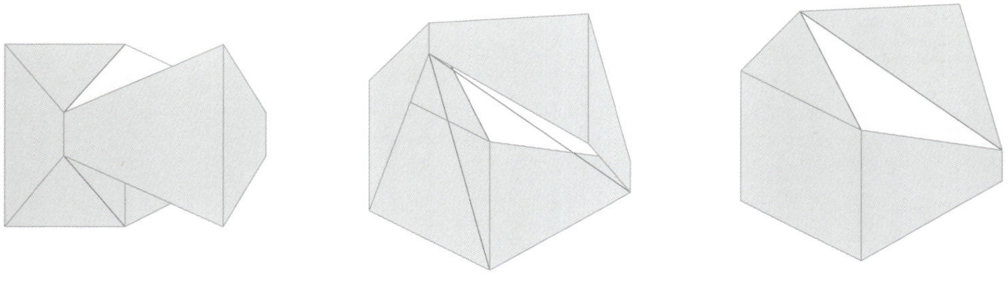

Brunner Exhibition Stand

布伦纳展厅

Project Location: Germany
Architectural Design: Ippolito Fleitz Group
Completion Year: 2011
Project Area: 70 sq.m.
Photography: Brunner Gmbh

项目地点：德国
建筑设计：Ippolito Fleitz Group 设计公司
竣工时间：2011 年
项目面积：70 平方米
摄　　影：Brunner Gmbh

An Attracting Architect 绚烂空间

Project Description

A brand new product highlight takes centre stage on the Brunner exhibition stand at the Milan Furniture Fair: twin, a monobloc plastic chair. Brunner enters a new price segment with the introduction of this exceedingly lightweight and reasonably priced chair. The company aims to establish a presence in the design product sector as well as attracting architects as a disseminating force.

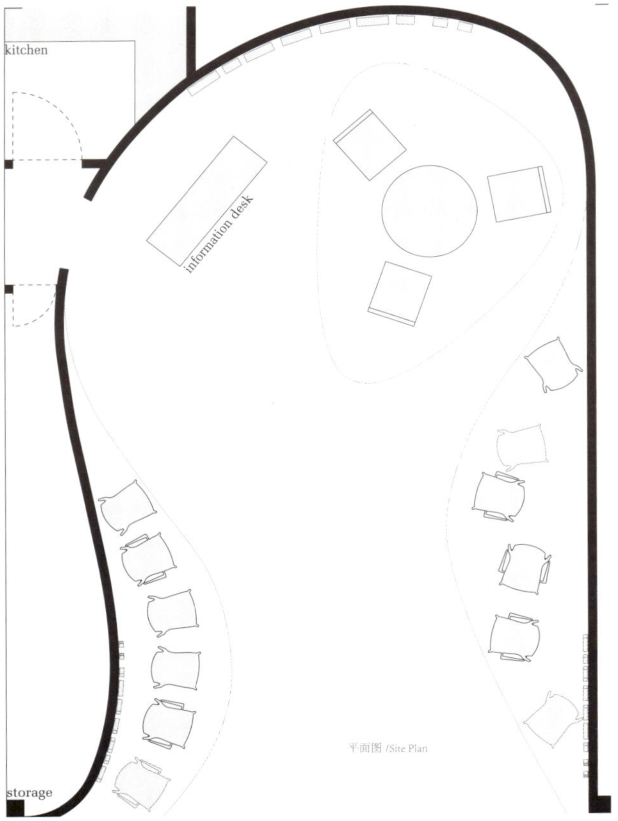
kitchen
information desk
storage
平面图 /Site Plan

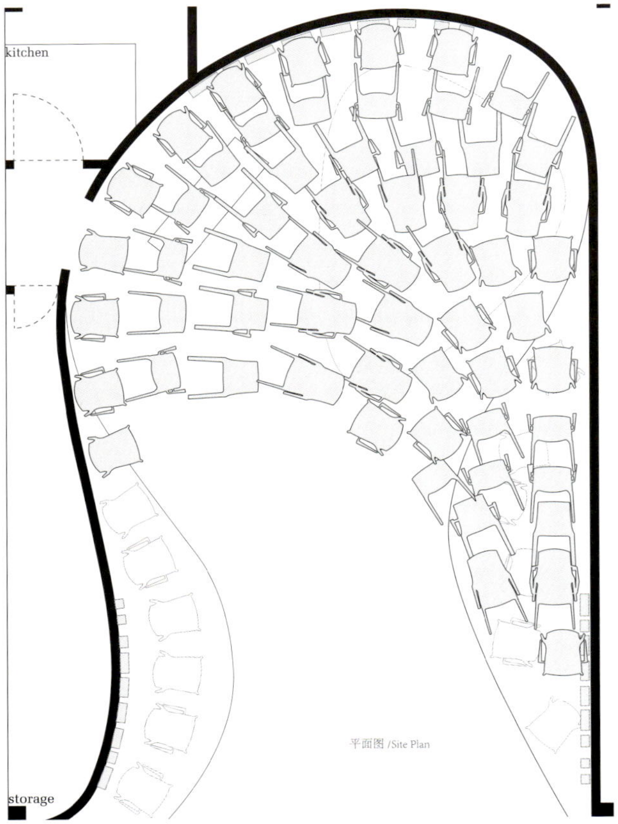
kitchen
storage
平面图 /Site Plan

brunner
nner

项目介绍

这是布伦纳家具品牌在意大利 2011 年国际移动家具展的展台，其设计极具活力，已超出单纯的展示设计，与公共装置艺术更接近了。一进门，首先看到的是许多具有动感造型的悬浮椅子，墙壁全部采用高反光的镜面贴片，映射出展厅周围的景象，给空间带来意想不到的效果。该项目采用六角形贴片，将零星的映射集合在一起，呈现出绚丽的色彩群。

brunner

图书在版编目（CIP）数据

城市装饰 / 香港理工国际出版社 主编 . – 武汉 : 华中科技大学出版社 , 2011.9

ISBN 978-7-5609-7278-7

Ⅰ . ①城… Ⅱ . ①香… Ⅲ . ①城市 – 环境设计 Ⅳ . ① TU-856

中国版本图书馆 CIP 数据核字 (2011) 第 166830 号

城市装饰 香港理工国际出版社 主编

出版发行：华中科技大学出版社（中国 · 武汉）

地　　址：武汉市武昌珞喻路1037号（邮编430074）

出 版 人：阮海洪

责任编辑：黎若君　唐丹丹

责任监印：张贵君

责任校对：段园园

装帧设计：吴美瑶

印　　刷：利丰雅高印刷（深圳）有限公司

开　　本：1016 mm × 1179 mm　1/16

印　　张：19.25

字　　数：153千字

版　　次：2011年9月第1版第1次印刷

定　　价：280.00元（US $43.99）

华中出版

投稿热线 :（010）64155588-8000　hzjztg@163.com

本书若有印装质量问题，请向出版社营销中心调换

全国免费服务热线：400-6679-118 竭诚为你服务